Textbook of Immunology

Second Edition

Seemi Farhat Basir
Professor and Former Head
Department of Biosciences
Jamia Millia Islamia
New Delhi

PHI Learning Private Limited
Delhi-110092
2015

₹ 250.00

TEXTBOOK OF IMMUNOLOGY, Second Edition
Seemi Farhat Basir

ISBN-978-81-203-4576-8

Third Printing (Second Edition) ... ... **August, 2015**

Published by Asoke K. Ghosh, PHI Learning Private Limited, Rimjhim House, 111, Patparganj Industrial Estate, Delhi-110092 and Printed by Mudrak, 30-A, Patparganj, Delhi-110091.

Contents

PREFACE

Immunology is the study that bridges medicine and basic sciences. It includes approaches from many fields like Biochemistry, Genetics, Cell Biology, Structural Biology and Molecular Biology. It is a broad branch of Biomedical Science that covers the study of all aspects of the immune system in all organisms, and deals with the physiological functioning of the immune system in both health and disease. It is a difficult task to bring Immunology to undergraduate students without overwhelming them or confusing them. This book has been written in a manner that makes it easy for the student to access knowledge about the topic of interest. It provides a broad overview of Immunology in a logical fashion. Hence, it is largely principles, facts, theories and findings that have been addressed, while deliberately leaving out the details of the experiments. As a teacher of Immunology, I have tried to present the subject in a simple and understandable manner. The aim of this book is to provide a single volume textbook that would satisfy the needs of both undergraduate and postgraduate students of Biosciences, Biotechnology and Biochemistry.

The decision to revise the first edition of the book was made after getting feedback from contemporaries and students. In the Second Edition a new chapter on *Immunity to Infectious Agents* is added and all the existing chapters have been upgraded.

The text has been divided into sixteen chapters covering different aspects of Immunology. It begins with an introduction about the immune system, and the organs and cells involved in the system. An idea about immunogens, antigens and antibodies follow letting onto the generation of diversity of antibodies. Topics like tolerance, hypersensitivity, transplantation and immunosuppression have also been dealt with. Efforts have been made to include tumor immunology, autoimmunity, immunodeficiencies, vaccines and cytokines. All the topics have been supplemented with figures and/or tables to make their understanding better and easier. In-text questions have been included at the end of each chapter to reinforce the understanding of the subject matter. I hope the readers will find this book very helpful.

I have enjoyed studying the concepts of immunology in *Immunology* by Kuby and *Immunology* by Tizard. I gratefully acknowledge these sources for many ideas and figures that I have adapted for this book.

My husband, Professor F.B. Khan, and children, Zarine and Mansur, gave me unstinting emotional and practical support. I am thankful to them for putting up with my academic activities.

I wish to thank my boys, Apoorv, Danish, Khubaib, late Yogesh and Zafar, who had been at my side during the good and specially the bad times that I had while writing this book. They dragged me to witness the various matches they played, whenever they thought I needed a break from my work. I also wish to acknowledge the help that Adeel and Hemajit provided to me whenever I got stuck at the computer.

Any constructive comments or suggestions for improving the contents of the book will be warmly appreciated.

Seemi Farhat Basir

1

INTRODUCTION TO THE IMMUNE SYSTEM

1.1 IMMUNE SYSTEM: AN OVERVIEW

The immune system is a collection of mechanisms within an organism that protects against disease by identifying and killing pathogens and tumour cells. The human body is equipped with an immune system, which is composed of lymphoid tissues and a variety of cells and soluble molecules. The immune system has the ability to respond to any antigen, such as a microbe or various macromolecules like proteins and polysaccharides, which are recognized as non-self or foreign, in order to protect our body from any potentially infectious organisms. The immune system is highly adaptable and generates a huge variety of cells and molecules that recognize and eliminate foreign invaders. The system recognizes and responds to the invaders to provide protection. It can distinguish foreign invaders from self-components and also detect minute chemical differences that distinguish two pathogens. It also recognizes altered host cells and those that might lead to cancers. Recognition of a pathogen by the immune system triggers an effector response that eliminates or neutralizes the invader.

Infectious microbes, like viruses, infect individual cells; others like many bacteria, divide extracellularly within tissues or the body cavities. These can cause disease, and if they multiply unchecked they eventually can kill the host. Most infections are short-lived in normal individuals and leave little permanent damage. This is due to the immune system that has evolved in vertebrates to protect them from invading pathogenic microorganisms and cancer. It is able to generate an enormous variety of cells and molecules capable of specifically recognizing and eliminating a limitless variety of foreign invaders.

The word **immunity**, which means *the state of protection from infectious diseases*, is derived from the Latin term *immunis* meaning 'exempt'.

There are two systems of immunity, innate and adaptive. These are the nonspecific and specific components respectively. The nonspecific component, **innate immunity** is a set of disease resistance mechanisms that are not specific to a particular pathogen. Moreover, it does not alter on repeated exposure to a given infectious agent. Phagocytic cells, such as macrophages, play an important role in many aspects of innate immunity. **Innate immunity** includes molecular and cellular mechanisms that are predeployed before an infection and poised to prevent or finish it. The recognition elements of the innate immune system distinguish between self and pathogens, but cannot distinguish small differences between foreign molecules.

In contrast, the specific component, **adaptive immunity**, displays a high degree of specificity as well as the remarkable property of 'memory'. It also improves with each successive encounter with the same pathogen. **Adaptive immunity** develops in response to infection and adapts to recognize, eliminate and then remember the invading pathogen. Adaptive immunity begins a few days after the initial infection and one consequence of this response is memory. If a same or closely related pathogen infects the body, memory cells provide the means for the adaptive immune system to make a rapid and highly effective attack on the invading pathogen.

Any immune response firstly involves recognition of the foreign material, and secondly mounting a reaction against it to eliminate it.

The immune system is able to recognize subtle chemical differences that distinguish one foreign pathogen from another. It is also able to discriminate between foreign molecules and the body's own cells and proteins. Once a foreign organism has been recognized, a variety of cells and molecules of the immune system participate in mounting an appropriate response, called an **effector response**, to eliminate or neutralize the organism. Later exposure to the same foreign organism induces a **memory response**, characterized by a more rapid and heightened immune response that serves to eliminate the pathogen and prevent disease.

1.2 HISTORY

Thucydides was the first person to make a reference to immunity when he described in 430 BC, a plague in Athens. He wrote that only those who had recovered from the plague could nurse the sick because they would not contact the disease a second time. The first attempts to induce immunity deliberately were performed by the Chinese and Turks in the fifteenth century while attempting to prevent smallpox. The dried crusts obtained from smallpox pustules were either inhaled or inserted into small cuts in the skin (a technique called **variolation**) in order to prevent this disease. In 1718, Lady Mary Wortley Montagu, the wife of the British ambassador in Constantinople, saw the positive effects of variolation on the native Turks and had the process performed on her own children. Some time later, an English physician Edward Jenner observed that milkmaids who had contacted a mild disease, cowpox, were subsequently immune to the much more severe smallpox. He reasoned that introducing fluid from cowpox pustule into people might protect them from smallpox. To test this idea, in 1798, Jenner inoculated an eight-year-old boy with fluid from a cowpox pustule and later intentionally infected the child with smallpox. In accordance to Jenner's expectations the child did not develop smallpox. Jenner's technique of inoculating with cowpox to protect against smallpox spread quickly throughout Europe, but it was another century before this technique was applied to other diseases.

The next major advance in immunology was the induction of immunity to cholera. Louis Pasteur had succeeded in growing the bacterium thought to cause fowl cholera in culture, and confirmed its role when chicken injected with the cultured bacterium developed fatal cholera. After returning from a vacation, he injected some chicken with the old bacterium culture. The chicken became ill, but to the surprise of Pasteur, they did not die and recovered. Pasteur then grew a fresh culture of the bacterium, intending to inject it in some fresh chicken. But his supply of chicken was limited, and so he used the previously injected chicken. Unexpectedly, the chicken were completely protected from the disease. Pasteur hypothesized and proved that aging had weakened the virulence of the pathogen and that such an attenuated strain might be administered to protect against the disease. Pasteur demonstrated that it was possible to weaken (attenuate) a pathogen and administer the weakened strain as a vaccine. In 1881, Pasteur first vaccinated one group of sheep with heat-attenuated anthrax bacteria (*Bacillus anthrasis*). He then challenged some vaccinated sheep and some unvaccinated sheep with the virulent culture of the anthrax bacillus. All the vaccinated sheep lived while all the unvaccinated animals died.

1885 saw Pasteur administer the first vaccine to a human being. Joseph Miller, a young boy who had been bitten by a rabid dog, was inoculated with attenuated rabies virus preparation. He lived and later became a custodian at the Pasteur Institute. Pasteur thus proved that vaccination worked, but did not know how.

In 1890, Emil von Behring and Shibasaburo demonstrated that serum from previously immunized animals to diphtheria could transfer the immunized state to unimmunized animals. This work fetched the Nobel Prize to von Behring in 1901. (Table 1.1 Nobel prizes for immunologic research). Research over the next ten years demonstrated that an active component from immune serum could neutralize toxins, precipitate toxins and agglutinate bacteria. A fraction of serum, then called gamma globulin and now called immunoglobulin, was shown to be responsible for all these activities through the efforts put in 1930s by Elvin Kabat. Because immunity was mediated by antibodies contained in body fluids (known at the time as humors), it was called **humoral immunity**. Whereas a vaccine causes active immunity in the host, transfer of antibody with a given specificity confers passive immunity. Before the discovery that a serum component could transfer immunity, Elie Metchnikoff in 1883, demonstrated that cells also contribute to the immune state of the animal. He observed that some white blood cells, which he termed phagocytes, ingested microorganisms and other foreign material. Since these phagocytes were found to be more active in immunized animals, Metchnikoff hypothesized that cells rather than serum components were the major effectors of immunity **(cellular or cell mediated immunity)**. It is now obvious that both humoral and cell mediated responses are required for a full immune response.

At the Rockefeller Institute, Merrill Chase in the 1940s succeeded in transferring immunity against tuberculosis organism by transferring white blood cells in guinea pigs. In the 1950s, the lymphocyte was identified as the cell responsible for both cellular and humoral immunity. Experiments pioneered by Bruce Glick at Mississippi State University indicted that T-lymphocytes mediated cellular immunity, and B-lymphocytes were involved in the humoral immunity. Around 1900, Jules Bordet had discovered that non-pathogenic substances, such as red blood cells from other species could serve as antigens.

Table 1.1 Nobel Prizes in Immunologic Research

Year	*Recipient*	*Country*	*Research*
1901	Emil von Behring	Germany	Serum antitoxins
1905	Robert Koch	Germany	Cellular immunity to tuberculosis
1908	Elie Metchnikoff Paul Ehrlich	Russia Germany	Role of phagocytosis and antitoxins in immunity
1913	Charles Richet	France	Anaphylaxis
1919	Jules Bordet	Belgium	Complement-mediated bacteriolysis
1930	Karl Landsteiner	United States	Discovery of human blood groups
1951	Max Theiler	South Africa	Developmenmt of yellow fever vaccine
1957	Daniel Bovet	Switzerland	Antihistamines
1960	Macfarlane Burnet Peter Medawar	Australia Great Britain	Discovery of acquired immune tolerance
1972	Rodney R. Porter Gerald M. Edelman	Great Britain Unites States	Chemical structure of antibodies
1977	Rosalyn R. Yalow	United States	Development of radioimmunoassay
1980	George Snell Jean Dausset Baruj Benaserraf	United States France United States	Major histocompatibility complex
1984	Cesar Milstein Georges E. Kohler Niels K. Jerne	Great Britain Germany Denmark	Monoclonal antibodies
1987	Susumu Tonegawa	Japan	Gene rearrangement in antibody production
1990	E Donnall Thomas Joseph Murray	United States United States	Transplantation immunology
1996	Peter C. Doherty Rolf M. Zinkernagel	Australia Switzerland	Role of MHC in antigen recognition by T-cells
2002	Sydney Brenner H Robert Horvitz J. E. Sulston	South Africa United States Great Britain	Genetic regulation of organ development and cell death
2005	Barry J. Marshall J. Robin Warren	Australia Australia	Bacterium Helicobacter pylori and its role in gastric and peptic ulcer disease.
2008	Luc Montagnier Francoise B. Sinoussi Haraltzur Hausen	France France Germany	Discovery of HIV as cause for AIDS HPV causes cervical cancer
2011	Bruce A. Beutler Jules A. Hoffmann Ralph M. Steinman	USA USA USA	Activation of innate immunity Discovery of dendritic cell and its role in adaptive immunity

In the 1950s, Neils Jerne, David Talmadge and F. Macfarlane Burnet gave a **clonal selection theory**. According to this theory, an individual lymphocyte expresses membrane receptors that are specific for a distinct antigen. This unique receptor specificity is determined before the lymphocyte is exposed to the antigen. When the antigen binds to the specific receptor it activates the cell causing it to proliferate into a clone of cells that have the same immunologic specificity as the parent cell.

Organisms that cause disease are called pathogens and their means of attacking the host is called pathogenesis. How effective immunity can be achieved against these pathogens relies heavily on the nature of the individual microorganism. Viruses require mammalian cell for their replication and therefore, an effective defence strategy might be to kill any cell that is infected with virus before the life cycle of the virus is completed. For organisms that replicate outside the host cells, prompt recognition of these invaders should be done by antibodies and soluble molecules, followed by cellular and molecular immune mechanisms to finish the pathogen.

1.3 INNATE IMMUNITY

This is the less specific component of immunity, and it provides the first line of defence against infection. It consists of the defences that are ready for immediate activation before attack by a pathogen. **Innate immunity (or natural or non-specific)** comprises of four types of defensive barriers: anatomic, physiologic, phagocytic, and inflammatory.

Anatomic barriers

The first barriers that the pathogen has to encounter are the anatomic barriers. These are the ones that protect a host and comprise of the **skin and the mucosal membrane** which include the mucosal epithelia that line the respiratory, gastrointestinal and urogenital tracts and insulate the body's interior from external pathogens. The *skin* and the *surface of mucous membranes* are the **anatomic barriers**. The skin acts as a mechanical barrier that retards the entry of microbes. The continuous renewal of epithelial cells, lining the skin surface and acidic environment (pH 3–5) too retards growth of microbes. The outer layer of the skin, the epidermis, consists of several layers of tightly packed epithelial cells. The outer epidermal layer has mostly dead cells filled with a waterproofing protein, keratin. The dermis is composed of connective tissue and has blood vessels, hair follicles sweat glands and sebaceous glands. These anatomical barriers also mount active biochemical defenses by synthesizing peptides and proteins with anti-microbial activity. Psoriasin is an active, small protein produced by human skin. It has potent antibacterial activity against *Escherichia coli.* Incubation of E. coli on human skin for 30 minutes specifically kills the bacteria. The alimentary, respiratory and urogenital tracts and the eyes are lined with mucous membranes that consists of an outer epithelial layer and an underlying connective tissue layer. Pathogens usually make their entry by penetrating these membranes. Hence, some non-specific defense mechanisms have to oppose these entries. Saliva, tears and mucous secretions wash away the potential invaders and also have antibacterial and antiviral substances. The conjunctiva, and the alimentary, respiratory and urinogenital tracts are lined by mucous membranes. The viscous mucus, which is secreted by the epithelial cells of the mucous membrane, entraps foreign microorganisms.

The mucous, secreted by the epithelial layer of mucous membrane entraps the incoming microorganisms. The mucous membrane of the lower respiratory tract is covered with cilia, whose movement propels the mucous-entrapped microorganisms from these tracts. There are antimicrobial compounds present in saliva and in the epithelia of the mouth. A mixture of acid and digestive enzymes are found in the stomach. Moreover, the pathogenic microbes have to compete with the normal flora that live on the mucosal surfaces, for nutrients and attachment sites. Sebaceous glands secrete sebum, which maintains the pH of the skin between 3 and 5, which proves to be a hindrance for the growth of invading microorganisms.

Physiological barriers

The **physiologic barriers** that contribute to innate immunity include *temperature, pH* and *various soluble factors.* The normal body temperature inhibits growth of some pathogens. Fever response inhibits the growth of some other pathogens. The acidic environment in the stomach kills most of the ingested pathogens. A variety of soluble factors also contribute to nonspecific immunity. Among them are the proteins—lysozyme, interferon and complement. Lysozyme cleaves bacterial cell wall, while interferon induces antiviral state in uninfected cells. Complement, which is a group of serum proteins, lyses microorganisms and facilitates phagocytosis.

Phagocytic or endocytic barriers

These are another type of nonspecific immune defense. Various cells internalize and break down foreign macromolecules (endocytosis). There are also some specialized cells like blood monocytes, neutrophils and tissue macrophages, that internalize, kill, and digest whole microorganisms (phagocytosis).

A variety of **soluble factors** also contribute to innate immunity. Certain soluble factors are produced at the site of infection or injury and act locally. They include antimicrobial peptides like defensins and cathelicidins, as well as interferons. The antimicrobial peptides consist from 6 to 59 amino acid residues and most of them are cationic. They kill a wide variety of bacteria and are mostly produced by neutrophils. Defensins kill microbes within a few minutes. Even the slow acting antimicrobial peptides take about half an hour to eliminate the invaders. They often work by disrupting the microbial membranes. However because of their questionable status regarding in vivo stability, toxicity, and efficiency when administered clinically, these antimicrobial peptides are not in clinical use. **Interferon** is a group of proteins produced by virus-infected cells. They can bind to nearby cells and induce a generalized antiviral state as they have got antiviral activity. **Lysozyme** is found in the mucous secretions and in tears, and is able to cleave the peptidoglycan layer of bacterial cell wall. Other soluble effectors are produced at distant sites and arrive at their target tissues via the blood stream. Complement proteins and acute phase proteins fall into this group. **Complement** is a group of serum proteins that circulate in an inactive state. Some specific and non-specific immunologic mechanisms can activate the complement, which then has the ability to damage the membranes of the pathogenic organisms, leading to their destruction and clearance. Changes in the concentration of several serum proteins are seen during the acute phase of the disease, which is the phase that precedes recovery or death. The serum proteins whose concentration rises or falls in the acute phase are called **Acute Phase Response proteins (APR proteins)**. The acute response phase is induced by signals that travel through the blood from the site of injury or infection. The liver is major site of APR protein synthesis and the proinflammatory cytokines IL-1, IL-6 and TNF-α are the major signals responsible for induction of acute phase proteins.

Collectins are proteins that may kill certain bacteria directly by disrupting their lipid membranes or by aggregating the bacteria to enhance their susceptibility to phagocytosis. Many molecules involved in innate immunity have the property of pattern recognition, which is the ability to recognize a given class of molecules. Certain types of molecules are unique to microbes and the ability to recognize and fight invaders showing such molecules is a strong feature of innate immunity. Molecules that have the property of pattern recognition may be soluble like lysozyme and the complement proteins or they may be cell-associated receptors. Innate immunity uses a variety of receptors to detect infection. One pattern recognition molecule is toll-like receptors. Toll is a transmembrane signal receptor protein; related molecules with roles in innate immunity are known as **Toll-Like Receptors (TLRs).** It was in 1998, when proof came from the work of Bruce Beutler that TLRs are part of the normal immune physiology.

Other pattern recognition molecules are present in the bloodstream and tissue fluids as soluble, circulating proteins or bound to the membranes of neutrophils macrophages and dendritic cells. Mannose Binding Lectin (MBL) and Complement Reactive Protein (CRP) are soluble pattern recognition receptors that bind to microbial surfaces, and promote phagocytosis or making the invader a target for lysis by complement. Another soluble receptor of the innate immune system is Liposaccharide-Binding Protein (LBP) that signals a response to lipopolysaccharide which is a component of the outer cell wall of the gram negative bacteria. NOD proteins (Nucleotide-binding Oligomerization Domain) are also a group of receptors that have a role in innate immunity. These are cytosolic proteins. NOD1 and NOD2 recognize products derived from bacterial peptidoglycans.

Inflammatory barriers

These come into play when there is some tissue damage caused by a wound or by an invading pathogenic microorganism. Tissue damage and infection induce leakage of vascular fluid, containing serum proteins with antibacterial activity, and influx of phagocytic cells into the affected area. **Other factors** include genetic factors (especially the gene complex known as the **major histocompatibility complex**) that control immune responsiveness and expression of histocompatibility antigens, as well as age. In the elderly and the very young, there is seen a decreased immune responsiveness.

When some microorganisms make their entry into an organism in spite of these innate defences, receptors of the innate immunity play an important role in detecting the infection and triggering an effective defence against it.

1.3.1 Cell Types of Innate System

Neutrophils

These are the first cells to migrate to the sites of infection. They are essential for the innate defence against bacteria and fungi. Neutrophils display several toll-like receptors on their surfaces. They detect the peptidoglycans of the gram positive bacteria through TLR2, and the TLR4 detects the lipopolysaccharide present on the walls of gram-negative bacteria. Binding and phagocytosis by neutrophils is greatly enhanced when microbes are marked (opsonized) by the attachment of antibody or complement components or both. In neutrophils, monocytes and macrophages two more antimicrobial devices, oxidative and non-oxidative attacks, contribute to a highly effective and prolonged defence. The oxidative arm employs Reactive Oxygen Species (ROS) and Reactive Nitrogen Species (RNS).

Macrophages

Resting **macrophages** are activated when TLRs on their surfaces recognize microbial components, such as lipopolysaccharides and peptidoglycans, and cytokine receptors detect cytokines released by other cells as part of inflammatory response. Activated macrophages exhibit greater phagocytic activity, increased ability to kill ingested microbes and also secrete mediators of inflammation. They also express higher levels of Class II MHC molecules, which present the antigen to T_h-cells. Activated macrophages express high levels of inducible nitric oxide synthase, an enzyme that produces Nitric Oxide (NO) from arginine. Nitric oxide has potent antimicrobial activity and can combine with superoxide to yield even more potent antimicrobial substances. These substances account for much of the antimicrobial activity of macrophages against bacteria, fungi, parasite worms and protozoa. Macrophages also coordinate other cells and tissues of immune and other systems, by secreting various cytokines like IL-1, IL-6 and TNF-α. They also produce complement proteins that promote inflammation and help in eliminating pathogens. The major site of synthesis of complement components is the liver, but macrophages in other cell types also produce them.

Natural killer (NK) cells

These cells provide the first line defence against many viral infections. NK-cells target and kill virus-infected cells which are potential sources of more virus particles. Some viral infections are completely cleared by the NK-cells while some infections need the help of adaptive immunity. In the latter cases the NK-cells hold the infection in check until help come in the form of adaptive immunity. NK-cells produce interferon-γ(INF-γ) and TNF-α, both of which are immunoregulatory cytokines and can stimulate maturation of dendritic cells. INF-γ activates macrophages and regulates development of T_h-cells.

Dendritic cells

These cells interact with both T_h- and T_C-cells and thus, provide a link between innate and adaptive immunity. They use TLRs to recognize pathogens and can generate the reactive oxygen species and the reactive nitrogen species as well as antimicrobial peptides.

1.4 ADAPTIVE IMMUNITY

Innate immune system is uniform in all members of a species. The specific component of the immune system, the **adaptive immunity** comes into play when there is a recognized antigenic challenge to the organism. It responds with a high degree of specificity as well as the property of memory. The innate and adaptive immune systems operate as a highly interactive and cooperative system, and produce a very effective combined activity. Adaptive immunity is capable of recognizing and selectively eliminating specific foreign microorganisms and molecules. It is not the same in all the members of a species and displays four characteristic features:

1. The antigenic specificity of the adaptive immune system allows it to distinguish subtle differences between antigens.
2. The immune system is capable to generate immense diversity in its recognition molecules, and thus, is able to recognize millions of unique structures on foreign molecules. The adaptive system can recognize a single type of organism and differentiate among those with minute genetic differences.

3. Once the adaptive system has recognized and responded to an antigen, it shows immunologic memory, and if encounters the same antigen a second time shows a heightened response.
4. The immune system responds only to foreign antigens showing its capacity of self-nonself recognition.

This is very essential, for the outcome of an inappropriate response to self-molecules which can be fatal. Table 1.2 compares innate and adaptive immunity.

Table 1.2 Comparison between Innate and Adaptive Immunity

Attribute	*Innate Immunity*	*Adaptive Immunity*
Response time	Minutes/hours	Days
Specificity	Specific for molecules and molecular patterns associated with pathogens	Highly specific; minor differences recognized with high specificity
Diversity	Limited number of germ line encoded receptors	Highly diverse; large number of receptors arising due to genetic recombination of receptor genes
Self/non-self discrimination	Very good	Good; Occasional failures result in autoimmune diseases
Memory response	None	Good memory response seen
Soluble components of blood or tissue fluids	Many antimicrobial peptides and proteins	Antibodies
Major cell types	Dendritic cells, phagocytes, natural killer cells	Antigen presenting cells, T-cells, B-cells

1.5 COOPERATION BETWEEN LYMPHOCYTES AND ANTIGEN-PRESENTING CELLS

An effective immune response involves two major cell types: the lymphocytes and the antigen-presenting cells. Lymphocytes are produced in the bone marrow through haematopoiesis; they then migrate to the lymphoid organs via the blood and the lymphatic systems. Since they have antigen-binding receptors on their surface, lymphocytes show specificity, diversity, memory and self-nonself recognition.

B-lymphocytes, on maturation, express a unique antigen binding receptor on its membrane, known as the B-cell receptor which is a membrane bound antibody molecule. When naive B-cells first encounter an antigen that matches its receptor, the binding of the antigen to the antibody causes the cell to divide rapidly; its progeny differentiates into **memory B-cells** and **effector B-cells** called **plasma cells**. Memory B-cells have the same membrane bound antibody as the parent B-cell and they have a longer life span than the naive cells. A single plasma cell can secrete thousands of antibody molecules per second, which are the main effector molecules of the humoral immunity.

T-cells arise in the bone marrow and mature in the thymus. Maturing T-cells express a unique antigen-binding molecule, the T-Cell Receptor (TCR) on its membrane. There are three different well defined subpopulations of T-cells: **T helper (T_h) cells** which have glycoprotein CD4 on their membrane surface, **T cytotoxic (T_C) cells** that have glycoprotein CD8 on their membrane surface and **T regulatory (T_{reg}) cells** that carry CD4 on their surface, but are distinguished from T_h- and T_C-cells by cell surface markers associated with its stage of activation.

T-cell receptors can recognize only that antigen which is bound to cell membrane proteins called **Major Histocompatibility Complex (MHC) molecules**. Class I MHC molecules are expressed by nearly all nucleated cells while the Class II MHC molecules are expressed by only **Antigen Presenting Cells (APCs)**. APCs include macrophages, B-lymphocytes and dendritic cells, and are distinguished by two characteristics: they express Class II molecules on their surface and they can produce cytokines that can activate T_h-cells. When a T_h-cell recognizes and interacts with antigen Class II molecule complex, the cell is activated and starts to produce various cytokines. The secreted cytokines activate B-cells, T_C-cells and macrophages, and other cells that participate in immune responses. Differences in the type of cytokines produced by the activated T_h-cells determine different patterns of immune response. Activation of humoral and cellular responses both require the activity of the cytokines produced. Antigen presenting cells first internalize the antigen by phagocytosis or endocytosis, and then display part of it on their membrane bound to a Class II MHC molecule. The T_h-cell interacts with the antigen-Class II molecule complex on the membrane of the antigen-presenting cell. The antigen-presenting cell then produces a signal that leads to the activation of the T_h-cell.

The elements of adaptive immunity, namely, lymphocytes and antibodies, are found only in vertebrates. However, some type of innate immunity appears to be characteristic of multicellular organisms, and immunity has been demonstrated in organisms as different as insects, earthworms and higher plants.

Innate and adaptive immunity do not operate totally independent of each other. These two systems, non-specific and specific immunity, form an interactive and supportive network that erects an effective and formidable barrier to infection. The activation of innate immune response produces signals that stimulate and direct subsequent adaptive immune responses. In addition, the adaptive immune response generates signals and components that stimulate and increase the effectiveness of innate immune responses.

All responses of the immune system are controlled by several **regulatory mechanisms** that maintain homeostasis and return the immune system to a state that existed before the antigenic stimulation. After a specific immune response to an actual infection or immunization, the host may become *immunized* against the stimulating antigen or develop a state of *tolerance* to that antigenic stimulus.

Although immune system is categorized mainly into two types, both the types are a function of one integrated immune system that is composed of cells and soluble molecules interacting and functional cooperativity to eliminate the stimulating antigen.

However, once the immune response is initiated, both responses function as an integrated defense system. The type of immune response (Natural or Specific) that may occur depends on

1. Recognition of foreign antigen.
2. Characteristics of the antigenic stimulus.
3. Site of infection (extracellular or intracellular).

The natural immune responses are considered to be non-specific because they

1. Do not require prior exposure to a foreign antigenic configuration to be effective.
2. Do not have ability to recognize a foreign antigenic structure.
3. Do not have a memory and are not enhanced by subsequent exposures to an antigenic structure.

The acquired immunity is initiated by an exposure to a specific antigen that has successfully evaded natural immune responses and is subsequently recognized by the immune system as foreign or non-self. The induction or initiation of immune responses depends on the ability of the immune system to differentiate between self and non-self antigens. This determines whether a specific immune response will occur or not.

Specific immune responses are some of the most powerful defense mechanisms available for inactivation or removal of a potentially immunogenic molecule before it can damage the host. These responses may occur during the first or second (and subsequent) encounters with a particular antigen. Two mechanisms by which the immune system specifically responds to a particular immunogen are humoral immune resonse and cell-mediated immune response.

1.6 HUMORAL IMMUNE RESPONSE AND CELL-MEDIATED IMMUNE RESPONSE

Humoral immunity can be conferred on a non-immune individual by transfer of serum antibodies from an immunized individual. Cell-mediated immunity can be transferred only by administration of T-cells from an immunized individual. The humoral branch sees to the interaction of B-cells with the antigen that results in their proliferation and differentiation into antibody-secreting plasma cells. Effector T-cells generated in response to antigen are responsible for the cell-mediated immunity (Figure 1.1).

These immune responses function as one integrated and interdependent response to an antigenic stimulus, they differ from one another in

1. Response to a different type of antigen and its site of infectivity.
2. Mechanisms that are triggered during the response.
3. Components of the immune system, such as effector cells and molecules involved in eliminating the antigenic stimulus.

1.7 CONFERRING SPECIFIC IMMUNITY: CONNECTIONS BETWEEN INNATE AND ADAPTIVE IMMUNITY

An immune response is first detected when the invader interacts with soluble or membrane-bound molecules of the host capable of discriminating between self and non-self. These molecular sensors recognize particular overall molecular patterns and are therefore known as Pattern Recognition Receptors (PRRs). Molecular patterns on pathogens are called Pathogen-Associated Molecular Patterns (PAMPs). The PAMPs recognized by PRRs include sugars, certain proteins, some lipid bearing molecules and some nucleic acid motifs. Innate recognition that is restricted to molecular patterns on microbes focuses the innate system on substances that cause infection rather than substances that are non-self such as artificial knees. In adaptive immunity, on the other hand,

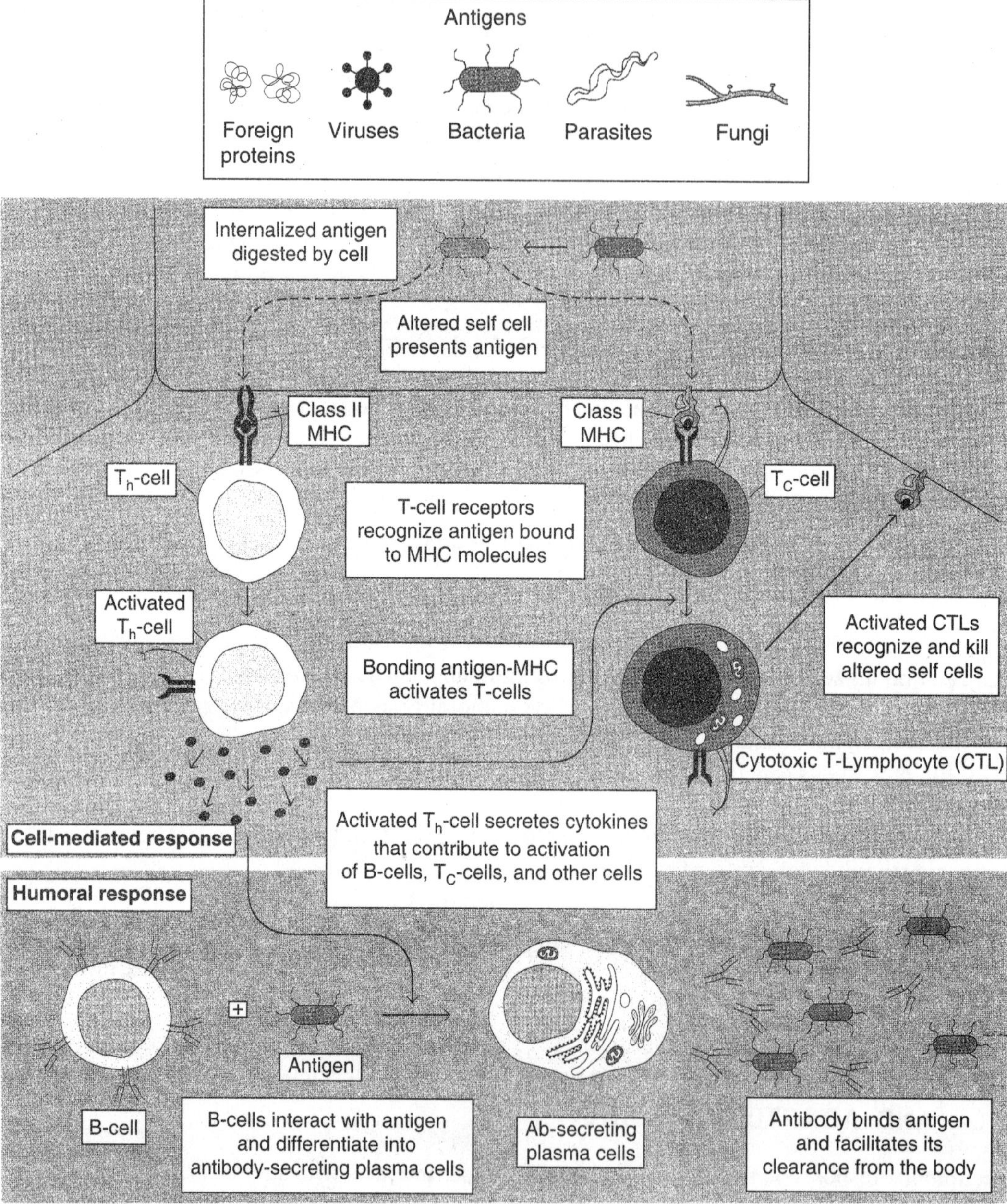

Figure 1.1 Effector T-cells in cell-mediated immunity.

antibodies and T-cell receptors recognize the finer differences between antigens. Detection of PAMPs by the soluble and membrane bound mediators of innate immunity brings multiple components of immunity into play. Soluble mediators include initiators of the complement system like MBL and CRP. A part of activated complement can form aggregates and make pores in the cell membrane of the targeted microbes thereby killing them. The invaded tissues dendritic cells and macrophages possess a variety of receptors, important among them being the toll-like receptors, which detect microbial products. So far 12 TLRs have been described in humans each of which reacts with a specific microbial product. Activated macrophages also secrete cytokines that communicate via cell receptors to induce specific cell activities. Presentation of the opsonized antigen to lymphocytes by the dendritic cells initiates an adaptive response.

Immunity to a particular antigen may be conferred on an individual either passively or actively. The resulting immune state is known as either **passive immunity** or **active immunity**.

Passive immunity

Passive immunity is conferred passively by administration of preformed cells or antibodies from another previously actively immunized individual. This type of immunity offers a rapid way of conferring resistance to a particular antigen without an actual exposure or an immune response of the recipient. This type of immunity is short lived, lasting only as long as the transferred cells or antibodies are present. Passive immunity is a very useful tool for quickly providing neutralizing antibodies (antitoxins) against toxins like tetanus and snake venoms, and as a preventive (antiviral) measure against hepatitis-B infection.

Active immunity

In active immunity, the individual becomes immunized by actually mounting an immune response to the stimulating antigen. Exposure of an individual to either an infectious or non-infectious foreign substance, triggers a humoral or cell-mediated immune response that results in an actively acquired specific immunity.

As the immune system encounters an antigen, one of the two events may occur in response to self or to a particular foreign antigen:

- Immune response
- Immunologic tolerance

Tolerance to self-antigens is naturally acquired. Tolerance to self-antigens is one of the most important features of the immune system because it prevents an immune response against the body's self-antigens and the development of autoimmunity. Self-tolerance is an acquired characteristic of the immune system by which self-reactive lymphocytes are destroyed or blocked from maturing or becoming activated as they encounter a self-antigen in the primary lymphoid organs.

However, it is also possible to artificially induce tolerance against a particular foreign antigen. Administering foreign antigens to very young animals, whose immature immune systems respond by becoming tolerant, may do this. Tolerance can also be provoked in adult animals by giving either extremely small or extremely large doses of antigen. Tolerance is specific for the inducing antigen and fades unless boosted by re-exposure to the same antigen. Self-tolerance is therefore, maintained by continuous exposure to normal body components throughout life.

1.8 CLONAL SELECTION AND EXPANSION

An immunocompetent mature animal has a large number of antigen-reactive clones of T-cells and B-cells. The antigenic specificity of each T- and B-lymphocyte is determined before its contact with antigen by random gene rearrangements during maturation, in the thymus or the bone marrow.

The role of antigen is important when it interacts with, and activate the mature antigenically committed T- and B-cells, resulting in the increase or expansion of population of cells with a given antigenic specificity. During this process of clonal selection, an antigen binds to a particular T- or B-cell and stimulates it to divide repeatedly into a clone of cells with the same antigenic specificity as the original parent cell (Figure 1.2).

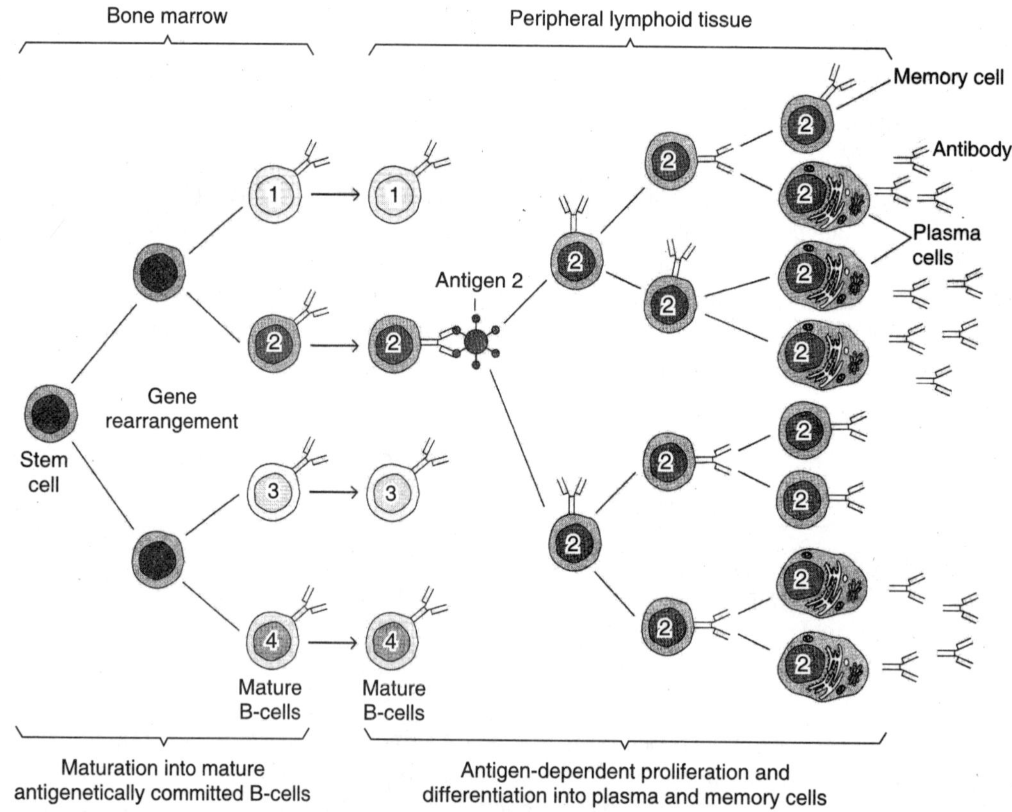

Figure 1.2 Clonal selection.

The Clonal Selection Theory was given by Medawar and Burnet, who got a Nobel Prize for the same in 1960. The postulates of the theory are as follows:

1. Each lymphocyte bears a single type of receptor of a unique specificity.
2. Interaction between a foreign molecule and a lymphocyte receptor capable of binding that molecule with high affinity leads to lymphocyte activation.
3. The differentiated effector cells derived from an activated lymphocyte will bear receptors of identical specificity to those of the parental cell from which that lymphocyte was derived.
4. Lymphocytes bearing receptors specific for self molecules are depleted at an early stage in lymphoid cell development and are, therefore, absent from the repertoire of mature lymphocytes.

1.9 FUNCTIONS OF IMMUNE SYSTEM

The immune system may also be defined in terms of its functions, which include:

Surveillance: This takes care of recognition of non-self foreign configuration, as well as recognition and destruction of mutant cells.

Defense: Which is initiation of an immune response to eliminate or inactivate a foreign antigen and microorganisms.

Regulation: Control of immune responses by regulatory mechanisms to maintain homeostasis.

Removal of damaged cells: This is done by apoptosis, which is a complex cell-death process that allows cells to die in a controlled fashion.

Immunity (immunologic memory): Resistance (acquired protection) following exposure to a stimulating agent.

Tolerance: Unresponsiveness towards certain antigens (mainly self antigens).

1.10 ORGANS OF THE IMMUNE SYSTEM

1.10.1 Sources of Lymphoid Cells

All the cells that participate in the immune responses originate from a population of **stem cells** that both (a) reproduce themselves, and (b) give rise to the specialized cell lines like neutrophils, macrophages, and lymphocytes.

Bone marrow is the source of all blood cells and is also a major source of antibodies. It consists of a meshwork of reticular tissue divided into two distinct compartments—a **hematopoietic** compartment and a **vascular** compartment. The hematopoietic areas contain precursors of all the blood cells, including macrophages and lymphocytes, and are enclosed by a layer of reticular cells. The vascular compartment of the bone marrow consists of blood sinuses lined by endothelial cells and crossed by reticular cells and macrophages. It is here that bacteria are trapped.

A number of morphologically and functionally diverse organs and tissues have various functions in the development of immune responses (Figure 1.3). These can be distinguished by their functions as **primary** and **secondary lymphoid organs.**

The thymus and the bone marrow are **the primary (or central) lymphoid organs**, where maturation of lymphocytes takes place. **The secondary (or peripheral) lymphoid organs** include the lymph nodes, the spleen, and various Mucosa-Associated Lymphoid Tissues (MALT) such as Gut-Associated Lymphoid Tissue (GALT). These organs provide sites for mature lymphocytes to interact with antigen. The mature lymphocytes circulate in the blood and the lymphatic system, a network of vessels that collect fluid that has escaped into the tissues from the capillaries of the circulatory system.

The immune system consists of cells and their secretory products, and the various lymphoid tissues and organs where these cells are organized and localized. The main components of the immune system, therefore, include:

(i) Lymphoid tissue (primary/secondary)
(ii) Cellular components (lymphocytes, accessory cells)
(iii) Soluble components/ mediators (cytokines, antibodies, complement components)

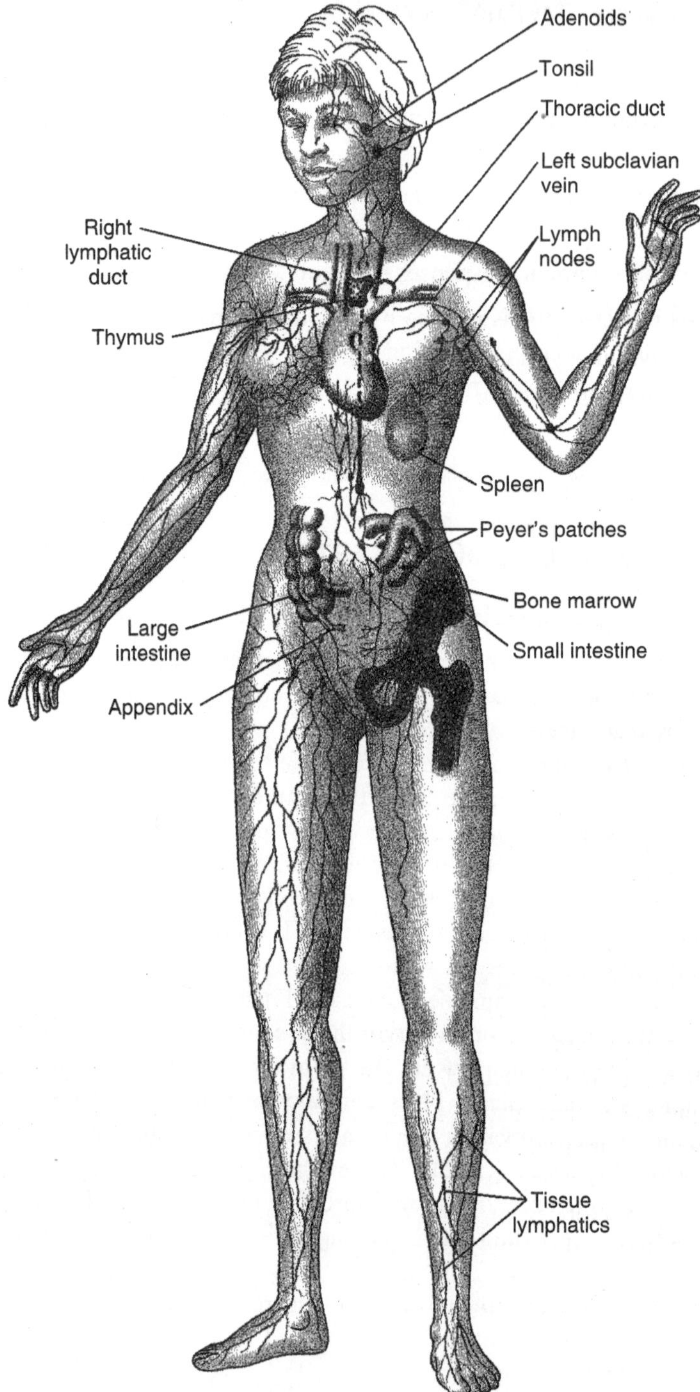

Figure 1.3 The human lymphoid system.

The primary functions of cells of the immune system and the various mediators that they secrete during an immune response is their participation in (1) the recognition of foreign antigen—the cognitive phase, (2) various responses of the immune system—the activation phase, and (3) resolution of the event by inactivating or eliminating the immune stimulus—effector phase.

1.10.2 Categorization of Lymphoid Tissue

Lymphoid tissue can be categorized into two divisions, according to their function as, Primary lymphoid tissue and Secondary lymphoid tissue.

Primary (generative) lymphoid tissue

The organs in which lymphocytes develop are called primary lymphoid organs. These include the bone marrow, the thymus, the bursa of Fabricius in birds, and some Peyer's patches in the intestine.

Immature lymphocytes generated in haematopoiesis mature and become committed to a particular antigenic specificity here. A lymphocyte becomes **immunocompetent** only after it has matured within a primary lymphoid organ. Thus the primary lymphoid organs provide sites where lymphocytes mature and become antigenically committed. T-lymphocytes mature within the thymus and B-lymphocytes mature within the bursa of Fabricius in birds and in the bone marrow of several, but not all, mammals. In both cases, a selection procedure eliminates immature lymphocytes that react with self-antigens. In addition, thymocytes that do not recognize self-MHC molecules are eliminated.

The primary lymphoid organs arise early in foetal life. They do not need the presence of foreign antigens to grow and are therefore, well-developed and fully functional in germ-free animals. These organs are seen to atrophy with age.

Bone marrow: All cells of the immune system are derived from an undifferentiated stem cell in the bone marrow, where they mature and differentiate by a process known as **haematopoiesis**. The bone marrow is a complex tissue and is the site of haematopoiesis. The haematopoetic cells formed in the bone marrow enter the bloodstream, which carries them out of the marrow and distributes the various cell types to the rest of the body.

In birds, a lymphoid organ called the bursa of Fabricius, a GALT tissue, is the primary site of B-cell maturation. In mammals like cattle and sheep, early foetal spleen and the ileac Peyer's patches act as the primary lymphoid tissue for the maturation, proliferation and diversification of B-cells. Rabbits also use GALT, such as appendix, as the primary lymphoid tissue.

In humans and mice, bone marrow is the site of B-cell origin and development.

In those animals whose immature B-cells proliferate and differentiate within the bone marrow, the stromal cells within the bone marrow interact directly with the B-cells and secrete various cytokines that are required for development. A selection process within the bone marrow eliminates B-cells with self-reactive antibody receptors.

The bone marrow thus acts as the source of all blood cells, including lymphocytes. It contains many macrophages that can remove foreign particles from blood. It also is the major source of antibodies.

The bone marrow consists of a meshwork of reticular tissue divided into (i) Haematopoietic compartment that contains precursors of all blood cells, including macrophages and lymphocytes, and (ii) Vascular compartment that consists of blood sinuses and macrophages, where bacteria are trapped.

From the lymphoid progenitors the immature B-cells proliferate and differentiate within the bone marrow, and stromal cells within the bone marrow interact with the B-cells and secrete various cytokines that are required for development. The selection process within the bone marrow eliminates B-cells that possess self-reactive antibody receptors.

Bursa of Fabricius: The bursa of Fabricius is an organ found in birds, but not in mammals. It is a hollow sac about 1 cm in diameter located just above the cloaca and connected to it by a duct. The bursa reaches its greatest size about one to two weeks after the chick has hatched and then it gradually atrophies. Inside the bursa, folds of epithelium extend into the lumen, and scattered through these folds are follicles of lymphoid cells. Each follicle is divided into a cortex and medulla. The cortex contains the lymphocytes, plasma cells and macrophages. The medullary epithelial cells are replaced by lymphocytes towards the centre of the medulla, so that the centre of the follicle appears to consist solely of lymphocytes. The bursa is a maturation and differentiation site for B-lymphocytes. Several hormones have been extracted from the bursa, the most important being **bursin,** a tripeptide (lys-his-glycylamide). Bursin activates B-cells, but not T-cells.

Peyer's patches: Peyer's patches are areas of lymphoid tissue located in the wall of the intestine (Figure1.4). In some species, such as the sheep, these Peyer's patches have a function similar to that of bursa. One type of Peyer's patches is found in the jejunum, and the other is found in the ileum and cecum. The ileocecal Peyer's patches consist of densely packed lymphoid follicles, each separated by a connective tissue sheath, and contain only B-cells. They reach their maximum

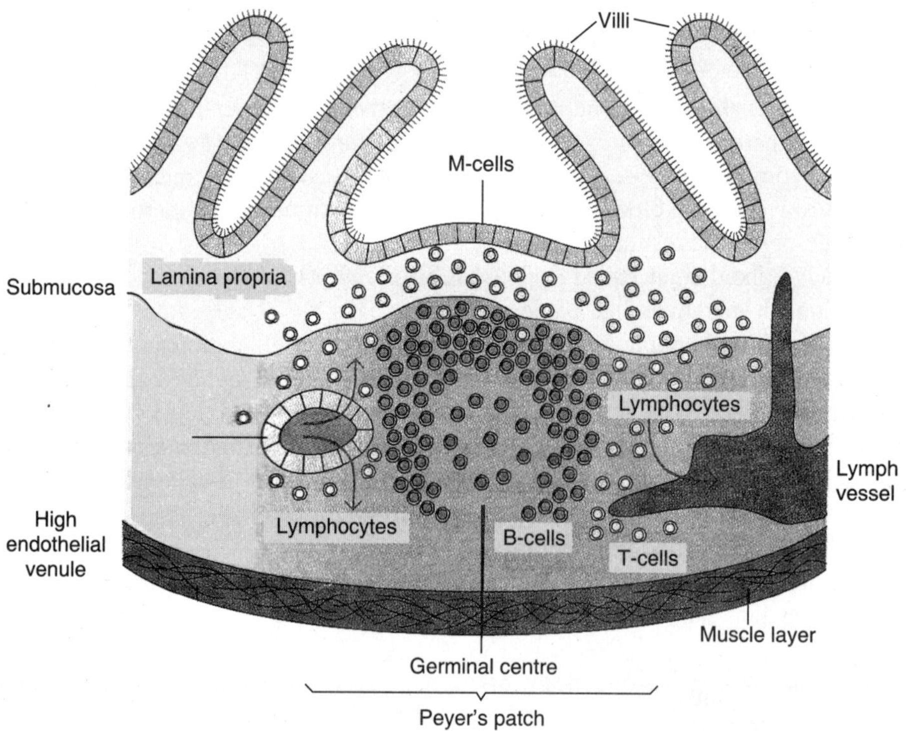

Figure 1.4 Cross-sectional diagram of the mucous membrane lining the intestine, showing a Peyer's patch.

size and maturity before birth, at a time when they are shielded from external antigens. Surgical removal of ileocecal Peyer's patches leads to a loss of circulating B-cells and a failure of antibody production. The ileocecal Peyer's patches are therefore, primary lymphoid organs and serve a function similar to the bursa of Fabricius.

Thymus: The thymus is a flat bilobed structure, located in the thorax just above the heart, between the lungs. Each lobe is surrounded by a capsule and divided into new lobules, which are separated from each other by strands of connective tissue called **trabeculae**. Each lobe has got an outer cortex that has a lot of immature thymocytes, and an inner medulla which has lesser number of thymocytes. Both these compartments have got a number of epithelial cells, dendritic cells and macrophages, which make up the framework of the organ and contribute to the growth and maturation of thymocytes.

Each lobe contains lymphoid cells (thymocytes) that form a tightly packed outer cortex and an inner medulla. There is a gradient of maturation of cells from cortex to medulla, as the cortex contains the immature and proliferating cells, while the medulla consists of more mature cells (Figure 1.5).

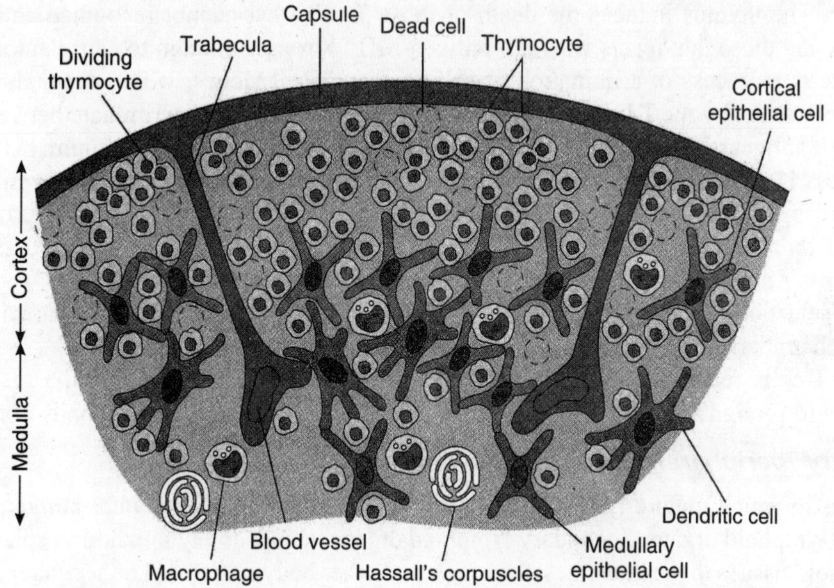

Figure 1.5 Diagrammatic cross-section of a portion of the thymus showing several lobules separated by trabeculae.

The thymus involutes with age, with only medullary remnants remaining. T-cell production however, continues throughout the life. All conditions associated with an increase in steroids, such as stress, promote thymic atrophy. Aging is accompanied by a decline in thymic function. The thymus reaches its maximal size at puberty and then atrophies, with a significant decrease in both cortical and medullary cells and an increase in the total fat content of the organ. Thymic function declines with age. The thymus reaches its maximum size at puberty and then atrophies. The average weight of the thymus is 30 gm in human infants, which declines to three gm in the elderly people. There is also a decline in T-cell output with age. At 35 years of age thymic generation of T-cells drops to 20 per cent of production in newborns, and by the age of 65 years the output falls to only 2 per cent of the newborn rate.

Entry of primitive cells from the bone marrow into the thymus is limited to the cells of lymphoid lineage that have been committed in the bone marrow to develop into T-lymphocytes. The lymphocytes leaving the bone marrow are attracted to the neonatal thymus by hormones secreted by thymic epithelial cells. The immature lymphocytes (called thymocytes) colonize the cortex and divide rapidly. Most of the new cells produced die within the cortex, but the survivors migrate to the medulla, mature, and eventually leave the thymus to colonize the **secondary lymphoid organs.** These thymus-derived lymphocytes are called **T-lymphocytes** or **T-cells.**

Thymic epithelial cells release several hormones, which are all polypeptides. They are not well defined and their immunological significance remains obscure. Their molecular weight range from 1 to 15-kDa and include thymosins, thymopoietins, thymic humoral factors, thymulin and thymostimulins. Thymic epithelial cells also produce interleukin-1.

The function of the thymus is to generate and select T-cells that protect the body from infection. During development the thymocytes develop a large variety of the cell receptors through gene rearrangement. This produces some T-cells with receptors capable of recognizing antigen-MHC complexes. The thymus induces the death of those T-cells that cannot recognize antigen-MHC complexes and those that reacts with self antigen-MHC strongly enough to cause autoimmunity.

There is a process of elimination of self-reactive T-cell clones, which occur during initial antigen recognition by the T-lymphocytes because of their ability to discriminate between the self and non-self antigens in the context of their MHC molecules, which they also acquire in the thymus. This process is known as **clonal selection**. Many of the cells entering into the thymus with the ability to recognize self-antigens do not survive as a result of this clonal selection process. *Only those cells that have acquired specific receptors for recognizing foreign antigens and cannot respond to self-antigens are stimulated to mature in the thymus.*

The selection procedure involves the destruction of thymocytes by apoptosis within **nurse cells**, which are large cells that can engulf up to 50 lymphocytes at a time.

The T-cells released into the circulation must participate in the immune responses by responding to foreign antigens. Moreover, they should not respond to normal body constituents.

Secondary (peripheral) lymphoid tissue

The organs in which mature lymphocytes reside and in which they encounter antigen are called secondary lymphoid organs. Secondary lymphoid organs include the lymph nodes, spleen, tonsils, and lymphoid tissues in the intestines, the lungs, and other body surfaces. The secondary lymphoid organs arise from the mesoderm late in foetal life and persist through adult life. They grow in response to antigenic stimulation and thus are poorly developed in germ-free animals. Removal of individual secondary lymphoid organs does not significantly affect the animal's ability to respond to antigen. The secondary lymphoid organs are rich in macrophages and dendritic cells that trap and process antigens, and in T- and B-lymphocytes which mediate the immune responses.

The secondary lymphoid tissues are present in two forms: (i) Encapsulated tissue (lymph nodes and spleen), (ii) Non-encapsulated tissue (cutaneous- and mucosa associated lymphoid tissue). The secondary lymphoid tissues are the primary sites of humoral and cell-mediated responses to foreign antigens.

The lymph nodes and spleen are the most highly organized of the secondary lymphoid organs; they have lymphoid follicles, and distinct areas of T-cell and B-cell activity that is surrounded by a fibrous capsule.

Some lymphoid tissue is organized into lymphoid follicles, which consists of aggregates of lymphoid and non-lymphoid cells surrounded by a network of draining lymphatic capillaries. Until activated by an antigen, a lymphoid follicle (called a **primary follicle**) comprises a network of follicular dendritic cells and small resting B-cells. After an antigenic challenge, a primary follicle becomes a larger **secondary follicle**—a ring of concentrically packed B-lymphocytes surrounding a **germinal centre** where there is a focus of proliferating B-lymphocytes and an area that contains non-dividing B-cells, and some helper T-cells interspersed with macrophages and follicular dendritic cells.

Following a period of division and mutation of the antigen activated B-cells in the germinal centres, there is a rigorous selection process in which more than 90 per cent of these B-cells die by apoptosis. The surviving B-cells differentiate into plasma cells and memory cells. The most highly organized of the secondary lymphoid organs are the lymph nodes and the spleen.

Less organized lymphoid tissue collectively called **Mucosa-Associated Lymphoid Tissues (MALT)** includes Peyer's patches (in the small intestine), the tonsils and the appendix as well as numerous lymphoid follicles in the mucus membranes lining the upper airways, bronchi, and genitourinary tract. MALT has the combined surface area of about 400 sq. metres. The membrane surfaces are the major sites of entry of pathogens. Epithelial cells of mucous membranes promote immune responses by delivering small samples of foreign antigen from the lumina of the respiratory, digestive and urogenital tracts to the MALT. Antigen transport is done by specialized M-cells. M-cells are flattened with no microvilli and having a deep invagination or pocket (in the basolateral plasma membrane) filled with a cluster of T-cells, B-cells and macrophages. The M-cells are located on inducted sites, which are small regions of a mucus membrane that lie over organized lymphoid follicles.

Lymph nodes: The lymph nodes are small aggregates of lymphoid tissue along the lymphatic channels throughout the body. Their structure is not fixed, but changes with the presence of antigen and regresses after its elimination. Lymph nodes are round or bean-shaped structures placed on lymphatic vessels in such a way that they can filter out any foreign material carried in the lymph. They consist of a fibrous reticular network filled with lymphocytes, macrophages and dendritic cells (Figure 1.6).

Lymph nodes are the sites where immune responses are mounted against antigens in lymph. They are bean shaped structures in a capsule. The outermost layer is the cortex that contains lymphocytes mostly B-cells, macrophages and follicular dendritic cells arranged in primary follicles. After antigenic stimulation, the primary follicles enlarged into secondary follicles each containing a germinal centre. Beneath the cortex is the paracortex, which is populated largely by T-cells, and interdigitating dendritic cells that migrate from tissues into the node. These dendritic cells express a lot of Class II MHC molecules, which are necessary for presenting antigen to T_h-cells. The innermost portion is the medulla, which has less number of cells of lymphoid lineage; there are plasma cells seen in this portion.

Morphologically, a lymph node is divided into three concentric regions. The outermost layer, the **cortex**, contains lymphocytes (mostly B-cells), macrophages and follicular dendritic cells arranged in primary follicles, which enlarge into secondary follicles after antigenic challenge. Beneath the cortex is the **paracortex**, which is largely populated by T-cells and interdigitating dendritic cells that have migrated from tissues to the node. The innermost layer of a lymph node, the **medulla**, is more sparsely populated with lymphoid cells; of those that are present, many are plasma cells actively secreting antibody molecules (Figure 1.7).

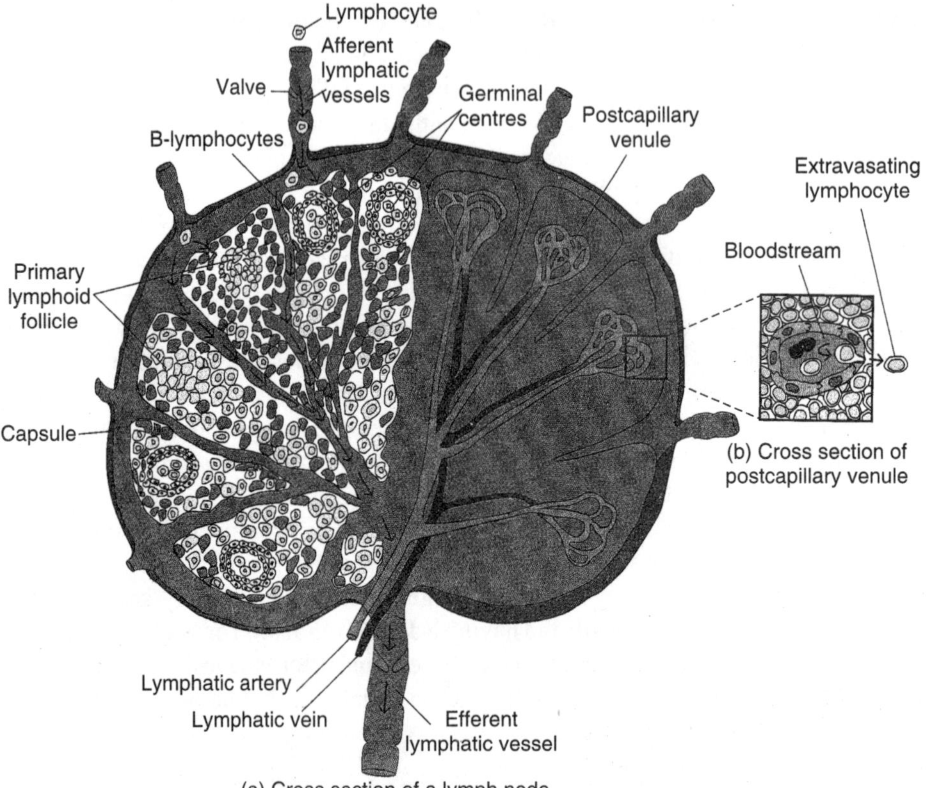

Figure 1.6 Structure of a lymph node.

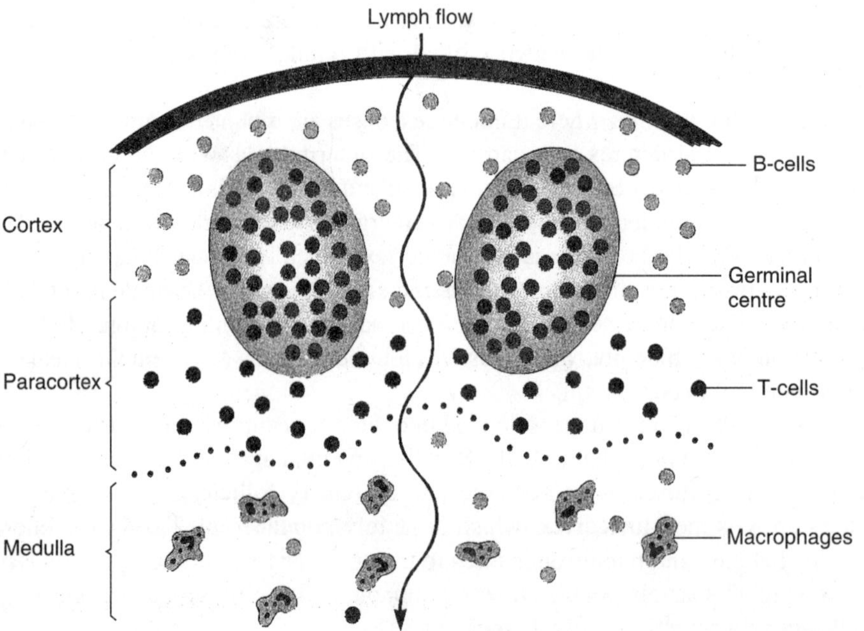

Figure 1.7 The distribution of B-cells, T-cells, and macrophages within a lymph node.

Afferent lymphatic vessels pierce the capsule of the lymph node at numerous sites and empties lymph into the subcapsular sinus. The lymph coming from tissues percolate inward through cortex, paracortex and medulla leading to entrapment of any bacteria or particulate matter. The lymph leaves through a single efferent lymphatic vessel and is rich in antibodies that are newly secreted by medullary plasma cells. This lymph has a 50 fold higher concentration of lymphocytes than the afferent lymph. Antigenic stimulation within the node increases the migration of blood-borne lymphocytes into the node and the node swells visibly. Factors released in the lymph nodes facilitate this increased migration of lymphocytes.

As antigen enters the lymph nodes through the lymphatic capillaries, the T- and B-lymphocytes present within the lymph nodes trigger an immune response and recognize it as foreign. These primary immune responses include the following:

- Production of soluble factors that create a cell-to-cell communication network (e.g. cytokines)
- Cell proliferation
- Development of effector cells
- Development of cytotoxic and helper T-cells
- Development of plasma cells that secrete antigen specific antibodies
- Development of T- or B-memory cells.

Memory cells developed during the primary immune response remain in the lymphoid tissue where they are ready to respond with a greater intensity and efficiency to the same antigenic stimulus, which they previously encountered during a primary immune response.

After antigenic stimulation, the cells within the primary follicles expand to form secondary follicles containing **germinal centres**. In the germinal centres, B-cells undergo somatic mutation, where the ability of the cells to bind to the antigen changes at random. The cells with increased ability to respond to an antigen leave the germinal centre to colonize other secondary lymphoid organs. Those cells with less ability to respond to an antigen undergo apoptosis and are removed by the macrophages. Germinal centres are also the site of immunoglobulin class switching.

Spleen: The spleen is an organ containing lymphoid tissue, lymphocytes, and accessory cells. The immune responses occurring in the spleen are in response to antigens that are blood-borne. The spleen has also the ability to 'clear' away foreign substances and senescent red blood cells by the process of phagocytosis by phagocytic cells present within the spleen. It also stores red blood cells and platelets and undertakes red blood cell formation in the foetus.

The spleen is a large, ovoid secondary lymphoid organ situated high in the left abdominal cavity. It specializes in filtering blood and trapping blood-borne antigens. It therefore, responds to systemic infections.

Lymphatic vessels do not supply spleen; instead the blood borne antigens and lymphocytes are carried into the spleen through the splenic artery. It is surrounded by a capsule that extends trabeculae into the interior. The spleen is divided into compartments of two types: (i) The red pulp, which consists of a network of sinusoids populated by macrophages and many red blood cells as well as erythrocytes. The defective RBCs are destroyed and removed in the red pulp. (ii) The white pulp which surrounds the branches of the splenic artery, forming a Periarteriolar Lymphoid Sheath (PALS). The white pulp is populated mainly by T-cells. There is a marginal zone located peripheral to PALS and is rich in B-cells that are organized into primary lymphoid follicles. Blood-borne antigens and lymphocytes enter the spleen through the splenic artery and are emptied into the marginal zone. Antigen is trapped here by dendritic cells, which then carry it to the PALS. The initial activation of B- and T-cells takes place in PALS and the primary follicles get changed to secondary follicles.

This organ is a major site for the production of antibodies and effector T-cells. It is divided into two compartments: the red pulp for the formation and storage of red blood cells, and the white pulp is where the immune responses occur. The white pulp has primary follicles containing B-cells. On antigenic stimulation, the cells within **germinal centres** are developed and the primary follicles expand to form secondary follicles. A layer of T-cells surrounds each follicle. This layer is known as the **mantle zone** (Figure 1.8).

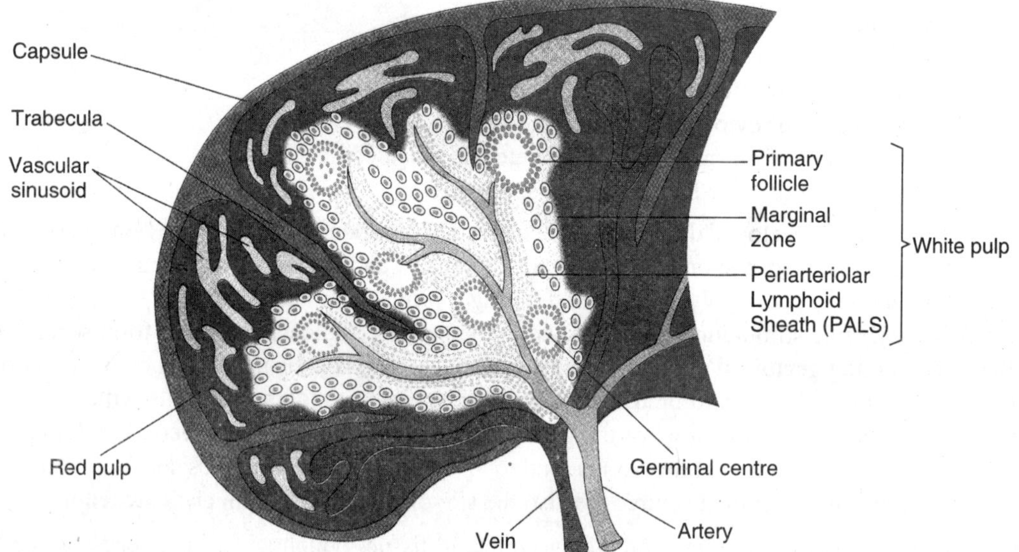

Figure 1.8 Structure of the spleen.

Adenoids, tonsils, Peyer's patches and appendix: They act as the temporary holding sites for recirculating lymphocytes. Some antigen processing also occurs here. The jejunal Peyer's patches have pear shaped follicles separated by inter-follicular tissue, and up to 30 per cent of their lymphocytes are T-cells. They persist throughout the animal's life and function as secondary lymphoid organs.

The secondary lymphoid organs function to capture antigen and to provide sites where lymphocytes interact with the antigen and undergo clonal proliferation and differentiate into effector cells. The lymphatic system drains the tissue spaces and interconnects many organized lymphoid tissues. Lymph nodes are specialized to trap antigen from regional tissue spaces, whereas spleen traps blood-borne antigens. Lymphoid tissue that is less organized is found in mucous membranes: these tissues include loose clusters of lymphoid follicles in the intestinal lamina propria and Peyer's patches found on the wall of the intestine. Cutaneous associated lymphoid tissue constitutes the most important tertiary lymphoid tissue.

Tertiary lymphoid tissue: This contains fewer lymphoid cells than secondary lymphoid organs and can import lymphocytes during an inflammatory response. Cutaneous Associated Lymphoid Tissue (CALT) comes under this category. Skin is a CALT and serves as an important anatomic barrier. The keratinocytes in the epidermal layer secrete a number of cytokines and may induce local inflammatory reactions. The keratinocytes can be induced to express Class II MHC molecules and then they are known as dendritic cells. Langerhans' cells are also a type of dendritic cells that are able to internalize antigen by phagocytosis or endocytosis. Intraepidermal lymphocytes are mostly $CD8^+$ T-cells. Dermal layer contains scattered $CD4^+$ and $CD8^+$ cells as well as macrophages too.

Lymphatic system: As blood circulates under pressure, its fluid component (plasma) seeps through the thin walls of the capillaries into the surrounding tissue. Much of this fluid, called **interstitial fluid,** returns to the blood through the capillary membranes. The remainder of the interstitial fluid, now called the **lymph,** flows from the spaces in connective tissue into a network of tiny open lymphatic capillaries and then into a series of progressively larger collecting vessels called **lymphatic vessels.**

When a foreign antigen gains entry into the tissues, it is picked up by the lymphatic system (which drains all the tissues of the body) and is carried to the various organized lymphoid tissues such as lymph nodes, which trap the foreign antigen. As lymph passes from tissues to lymphatic vessels, it becomes progressively enriched in lymphocytes (Figure 1.9). Thus, the lymphatic system also serves as a means of transporting lymphocytes and antigen from the connective tissue to organized lymphoid tissues where the lymphocytes may interact with the trapped antigen and undergo activation.

Haematopoiesis: All blood cells are derived from Haematopoietic Stem Cells (HSCs). There are two characteristics that these HSCs possess. Firstly, they can differentiate into other cell types, and secondly, they have the property of self-renewal and thus, maintain their population by cell division. Haematopoeisis is the process of formation of red and white blood cells, which begins in the embryonic sac in the first few weeks of development. The yolk-sac stem cells differentiate into primitive erythroid cells that contain embryonic haemoglobin. During the third month of gestation the HSCs migrate from the yolk sac to the foetal liver, from where they move to the spleen. The foetal liver and the spleen play a major role in haematopoiesis from the third to the seventh month of gestation. After the seventh month, the bone marrow becomes the major site of haematopoiesis, where differentiation of HSCs occurs. By the time of birth, there is very little or no haematopoiesis in the liver and spleen.

HSCs are pluripotent cells, i.e., they are able to differentiate in various ways and can generate erythrocytes, granulocytes, monocytes, mast cells and lymphocytes. The number of HSCs is quite low. There is one HSC for every 5×10^4 cells in bone marrow. HSCs are maintained at stable levels throughout adult life. Whenever there is an increased demand for haematopoiesis, the HSCs show enormous proliferative capacity. Early in haematopoiesis, a pluripotent stem cell gives rise to (i) a common lymphoid progenitor cell or (ii) a myeloid stem cell (Figure 1.10). The types and amounts of growth factors in the microenvironment of a particular stem cell control its differentiation. The progenitor cells loose their ability for self-renewal and are committed to a particular cell lineage. The common progenitor cells give rise to B-, T-, and NK-cells. The myeloid stem cells generate progenitors of RBCs, erythrocytes, neutrophils, basophils, monocytes, and mast cells.

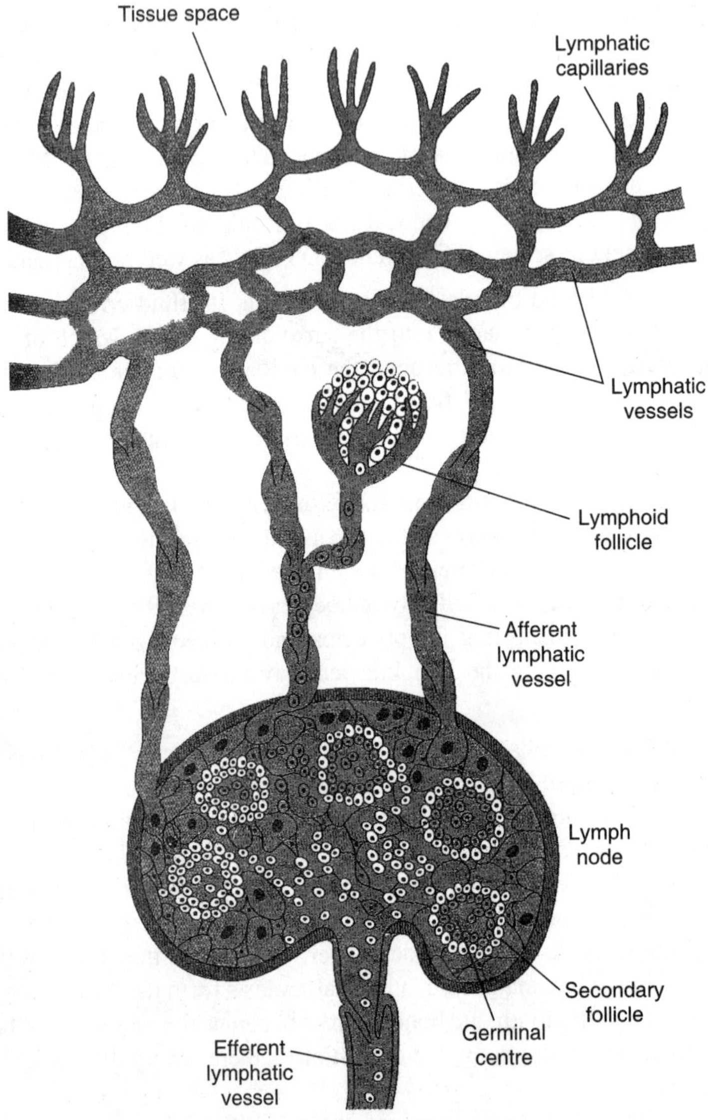

Figure 1.9 Lymphatic vessels.

The progenitor commitment depends upon the acquisition of responsiveness to particular growth factors and cytokines.

In adult bone marrow, the haematopoietic cells grow and mature on a meshwork of stromal cells, which are non-haematopoietic cells that support the growth and differentiation of haematopoietic cells. The stromal cells include fat cells, endothelial cells, fibroblasts and macrophages. They influence the differentiation of haematopoietic cells by providing a haematopoiesis inducing microenvironment, which consists of a cellular matrix, and growth factors that are either membrane bound or diffusible.

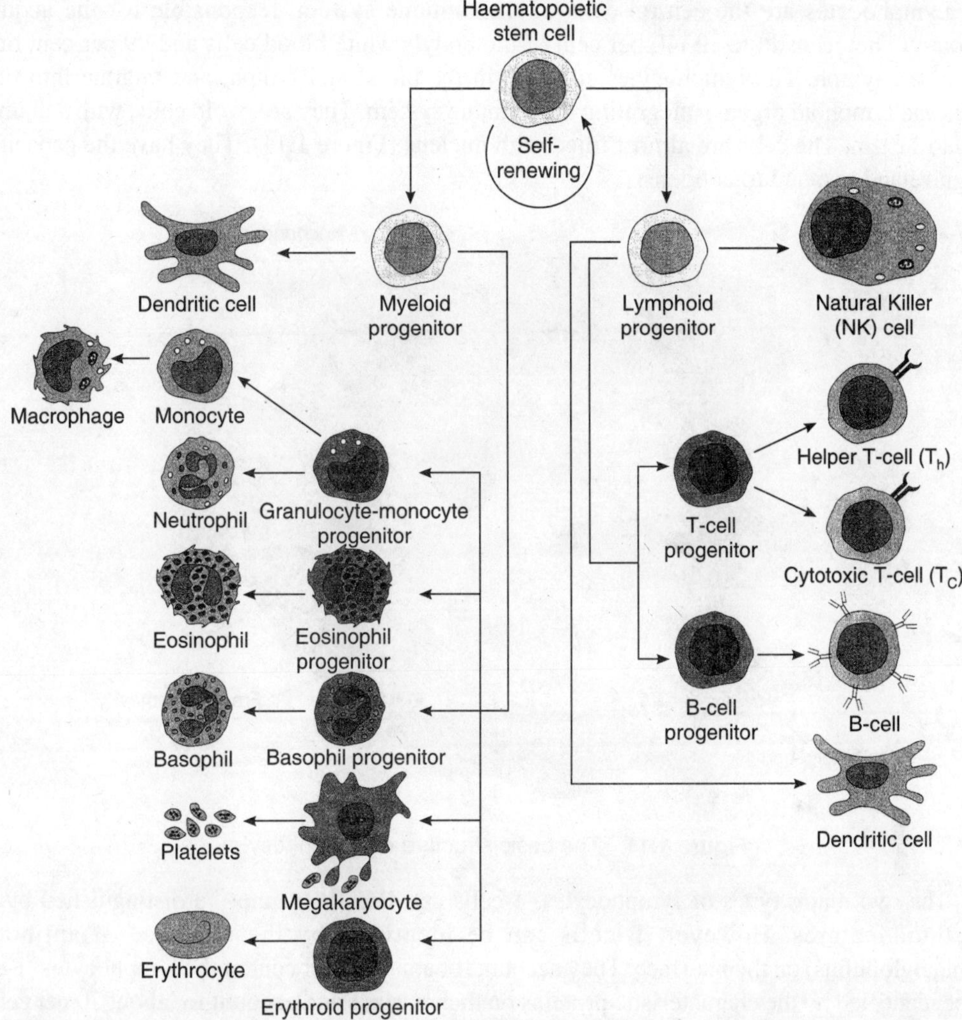

Figure 1.10 Haematopoiesis. Self-renewing haematopoeitic stem cells give rise to lymphoid and myloid progenitors.

1.10.3 Cells of the Immune System

Two major groups of cells are involved in an immune response. They are the lymphocytes and the antigen presenting cells.

Lymphocytes

Immune responses are produced primarily by leucocytes, which are of several types. The lymphocytes originate in the bone marrow and then leave the site of origin to circulate in the blood and lymphatic systems before residing in various lymphoid organs. Lymphocytes produce and display antigen binding cell surface receptors. They mediate the immunologic attributes of specificity, diversity, memory, and self-nonself recognition.

Lymphocytes are the central cells of the immune system, responsible for the acquired immunity. They constitute 20–40 per cent of the body's white blood cells and 99 per cent of the cells in the lymph. The lymphocytes circulate in the blood and lymph, and migrate into tissue spaces and lymphoid organs, integrating the immune system. They are ovoid cells, with a diameter of up to 12 μm. The cells are almost filled with nucleus (Figure 1.11). They have the capacity to recognize and respond to antigens.

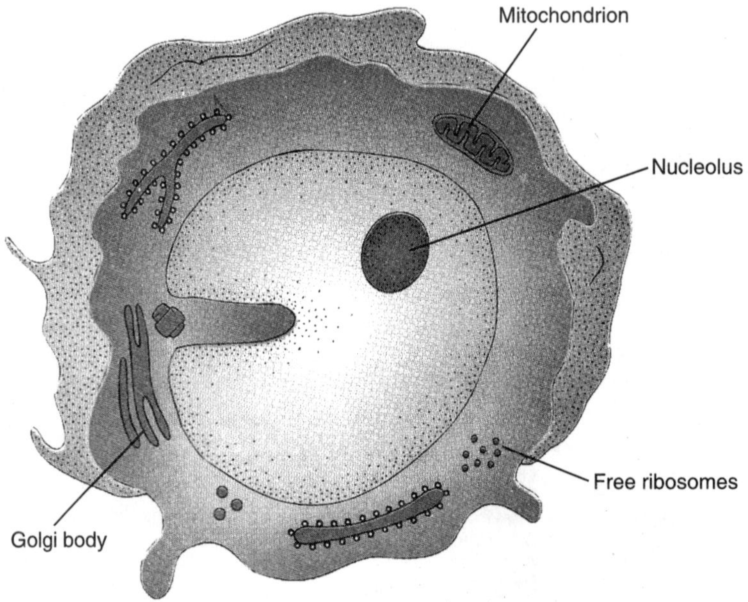

Figure 1.11 The basic structure of a lymphocyte.

The two major types of lymphocytes, T-cells and B-cells, cannot be distinguished by any structural features. However, B-cells can be identified by the presence of antibodies (immunoglobulins) on their surface. They account for about 20 per cent of the lymphocytes. T-cells can be identified by the characteristic proteins on their surface and account for about 70 per cent of the lymphocytes. The third type of lymphocytes are the null cells that do not express the surface markers typical of the T- and B-cells and make up less than 5 per cent of the lymphocytes.

B- and T-cells differ from each other in their origin, surface markers, circulating patterns, and mode and consequences of their interaction with antigens. The ratio of T : B-cells in the lymph nodes is 5 : 2 while the same is 1 : 2 in the spleen and the Peyer's patches.

T-lymphocytes: They arise in the bone marrow, but migrate to mature in the thymus and are critically involved in both humoral and cell-mediated immune responses. The quality and quantity of life may depend on the proper number of healthy T-cells. Among the pluripotent stem cells in the bone marrow are progenitor T-cells, which migrate to the thymus as thymocytes. Some T-cell markers are given here and also shown in Figure 1.12.

- Thy-1 or θ—this T-cell marker in mice distinguishes T-cells from B-cells. Its equivalent in humans is known as CD5.
- CD2—this is a glycoprotein. The protein part has the capability of binding sheep RBC.

- CD3—is a set of five proteins (γ, δ, ε, ζ and p-21), that form a complex. This cluster of differentiation is associated with antigen- specific receptors.
- CD4—is a glycoprotein of the T_h1 and T_h2 cells. The helper effect is through lymphokines (IL-4, IL-5 and IL-6) that the T_h2 cells secrete. The T_h1 cells secrete γ-interferon, which promotes delayed type of hypersensitivity. In human beings, CD4 is the receptor for HIV.
- CD8—$CD8^+$ T-cells are cytotoxic in nature and are called T killer cells (T_c- or T_k-cells).

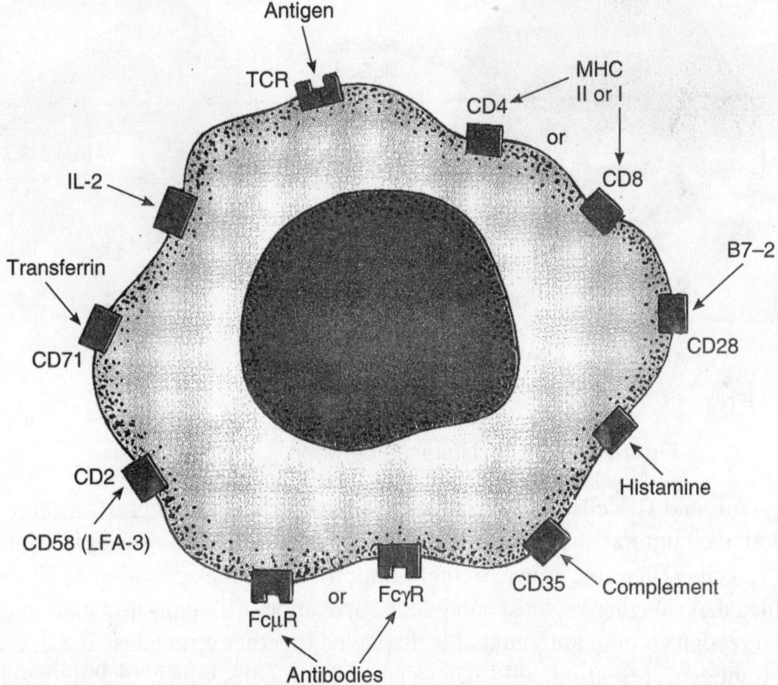

Figure 1.12 The major receptors found on the surface of T-lymphocytes.

T-cells that have CD4 on their surface are restricted to recognition of antigen bound to Class II MHC molecules, whereas T-cells expressing CD8 glycoprotein are restricted to recognition of antigen bound to Class I MHC molecules.

Mature T-cells express a unique antigen-binding molecule, the T-Cell Receptor (TCR) on their membranes. There are two well-defined sub-populations of T-cells: T Helper (T_h) and T cytotoxic (T_C) cells. T_h-cells have CD4 membrane glycoprotein on their surface, while T_C cells have CD8 membrane glycoprotein. **The ratio of $CD4^+$ to $CD8^+$ T-cells is about 2 : 1 in normal human peripheral blood.** Most T-cells receptors can recognize only antigen that is bound to cell membrane proteins called **Major Histocompatibility Complex (MHC)** molecules.

T-cells are further classified into the following subclasses:

T_h-cells. These are the T helper cells, and provide assistance of immune function by other lymphocytes. They help the B-cells to produce antibodies. These cells are of two types, namely, T_h1 and T_h2 (Figure 1.13).

T_s-cells. They secrete soluble factors that inhibit specific T_h-cells, and selectively shut off transcription of the Ig heavy and light chain genes.

T_C-cells. They are responsible for cytolysis and cell death of targets.

T_D-cells. They recruit and regulate a variety of non-specific blood cells and macrophages in the expression of delayed type of hypersensitivity.

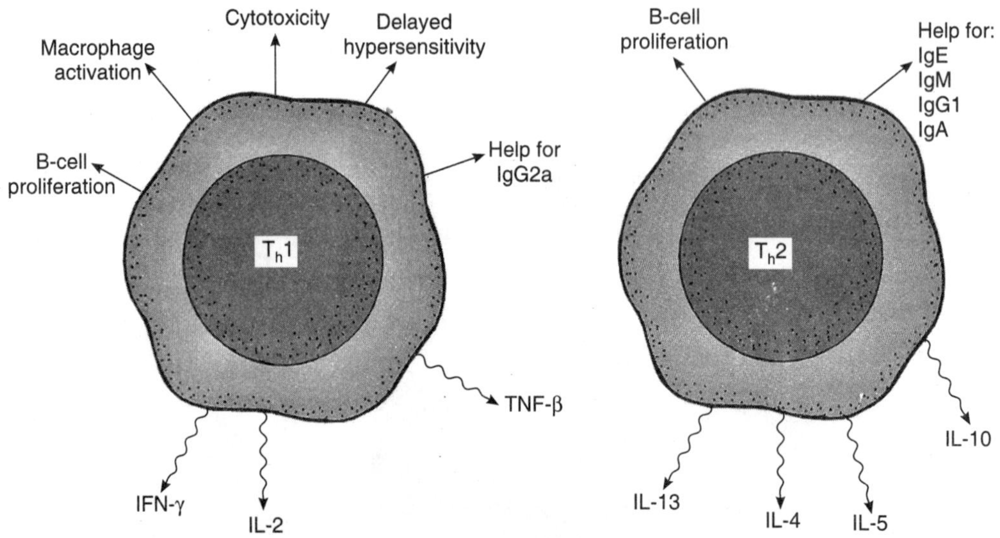

Figure 1.13 The difference between T_h1 and T_h2 cells.

Both T_C-cells and T_D-cells are directly involved in the rejection of grafts that are incompatible to the host. Both the humoral and cell mediated branches of the immune system requires cytokines produced by T_h-cells. Therefore it is extremely important that the activation of T_h-cells should be carefully regulated. A safeguard against unregulated activation of T_h-cells is that the antigen receptors of T_h-cells can recognize only antigen that is displayed together with Class II MHC molecules on the surface of antigen presenting cells. These specialized cells which include macrophages, the lymphocytes and dendritic cells have two properties: (i) the express Class II MHC molecules on their membranes, and (ii) they can produce cytokines that cause T_h-cells to become activated.

Recognition of antigen-MHC complexes by a T_C-cell triggers its proliferation and differentiation into an effector cell called a Cytotoxic T-Lymphocyte (CTL) or into a memory cell. T regulatory cells are identified by the presence of both CD4 and CD25 on their membranes. T_{reg}-cells suppress immune responses.

Progenitor T-cells from the bone marrow enter the thymus and rearrange their TCR genes. The earliest thymocytes lack detectable CD4 and CD8 and are called double negative cells. During development, the majority of double negative thymocytes develop into $CD4^+$ CD 8^- or $CD4^-$ $CD8^+$ T-cells. Positive selection in the thymus eliminates T-cells unable to recognize self-MHC. Negative selection eliminates thymocytes bearing high-affinity receptors for self-MHC molecules alone or self-antigen plus self-MHC and produces self-tolerance. T-cell activation is initiated by interaction of the TCR-CD3 complex with a peptide-MHC on an antigen-presenting cell; it also requires the activity of co-receptors CD4 and CD8. TCR engagement with antigenic peptide-MHC may induce either activation or clonal anergy.

B-lymphocytes: The B-lymphocytes are derived from the pluripotent stem cells in the bone marrow and on release from this site start expressing a unique antigen binding receptor or B-cell receptor which is a membrane-bound antibody molecule.

Bone marrow is the major site of maturation of B-lymphocytes. Mature B-cells are distinguished from other lymphocytes and all other cells by the synthesis and display of membrane-bound immunoglobulin (antibody) molecules which act as receptors for antigen.

They mature in the bone marrow itself and then migrate to the secondary lymphoid organs. The antigen dependent maturation occurs in the spleen in mammals and in the foetal liver too. The pre B-cells are large lymphoid cells, which have Ig markers of the M type. The B-cells have about 100,000 Ig molecules per cell. Some B-cell markers are given here and also shown in Figure 1.14.

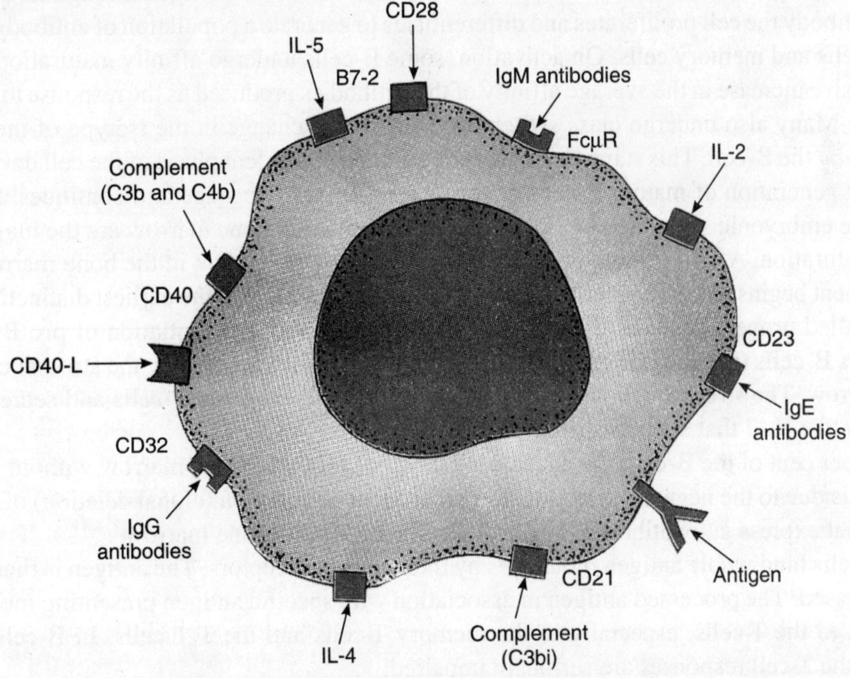

Figure 1.14 The major receptors found on the surface of B-lymphocytes.

Igs. These are surface membrane bound immunoglobulins that serve as antigen-specific receptors.

Class II MHC molecules. Permit the B-cell to function as an antigen-presenting cell.

CD40. This is a molecule that interacts with CD40 ligand on the surface of T helper cells. This interaction is critical for the survival of antigen stimulated B-cells and their development into plasma or memory cells.

$C3b_i$ receptor. This is a receptor for a fragment of complement and also for the Epstein Barr virus, which infects B-cells and converts them to tumour cells.

Cytokine receptors. Especially for interleukin-1.

The B-cells also have receptors for substances produced by the T-cells.

Receptors for mitogens.

When a naive B-cell first encounters the antigen that matches its membrane-bound antibody, the binding of the antigen to the antibody causes the cell to divide rapidly; its progeny differentiates into **memory B-cells** and effective B-cells called the **plasma cells.** Memory B-cells have a longer lifespan than naive cells and they express the same membrane-bound antibody as their parent B-cell. Plasma cells produce the antibody in a form that can be secreted and have little or no membrane-bound antibody. Plasma cells are end-stage cells, do not divide and many die within one or two weeks.

Immature B-cells bearing IgM on its membrane leaves the bone marrow and matures to express both membrane-bound IgM and IgD with a single antigenic specificity. These naive B-cells circulate in the blood and lymph and are carried to the secondary lymphoid organs specially the spleen and lymph nodes. If a B-cell is activated by interaction with antigen through its membrane-bound antibody the cell proliferates and differentiates to generate a population of antibody secreting plasma cells and memory cells. On activation, some B-cells undergo affinity maturation, which is a progressive increase in the average affinity of the antibodies produced as the response to activation proceeds. Many also undergo class switching which is the change in the isotype of the antibody produced by the B-cell. This stage constitutes the antigen dependent phase of the cell development.

The generation of mature B-cells occurs during embryonic stage and continues throughout life. In the embryonic stage the yolk sac, foetal liver and foetal bone marrow are the major sites of B-cell maturation. After birth the generation of mature B-cells occurs in the bone marrow. B-cell development begins as the lymphoid precursor cells differentiate into the earliest distinctive lineage B-cell, called progenitor B-cell (pro B-cell). Proliferation and differentiation of pro B-cells into precursors B-cells (pre B-cells) require the microenvironment provided by the stromal cells of the bone marrow. The stromal cells interact directly with pro B- and pre B-cells and secrete various cytokines like IL-7 that support the developmental process.

90 per cent of the B-cells produced each day die within the bone marrow without leaving it. This loss is due to the negative selection and subsequent elimination (clonal deletion) of immature B-cells that express autoantibodies against cell antigens in the bone marrow.

B-cells bind whole antigen molecules by their antigen receptors. The antigen is then ingested and processed. The processed antigen in association with specific antigen presenting molecules, is presented to the T-cells, especially to the memory T-cells and the T_h1 cells. In B-cell deprived animals, the T-cell responses are seriously impaired.

Cap formation. There is a formation of a polar cap by all the surface markers upon the interaction of antigen with the B-cell. This is followed by pinocytosis of the cap formed in which all the surface markers are destroyed. Then there is a re-expression of the surface markers after about 8 hours. This implies that there is a dynamic turnover in the surface markers. The capping and the internalization of the surface markers is a key step towards activation of lymphocytes. It is also possible that the cross linking of the surface receptors triggers cell proliferation and differentiation.

Plasma cells: Plasma cells are the fully differentiated antibody-synthesizing cells. These cells have an eccentric nucleus, radially arranged chromatin, numerous mitochondria, abundant RER (Rough Endoplasmic Reticulum), and prominent Golgi apparatus. The plasma cells do not have surface Ig markers. However, they do possess a unique Plasma Cell antigen (PC). The antibodies produced by a single plasma cell may be of any kind, but the antigen binding specificity is always identical to the original B-cell precursor.

Blast transformation. Upon incubation with an antigen, the small lymphocytes undergo what is known as blast transformation. The cells enlarge and their nucleolus swells. There is an increment in the polysomes, endoplasmic reticulum and microtubules. The rate of macromolecule synthesis also goes up. The binding of anti-Ig antibody to receptors of B-cells, and the binding of certain mitogenic plant proteins to the surface of B- and T-cells also initiate this blast transformation. This implies that cross-linking of appropriate surface macromolecules sets off a membrane perturbation that stimulates the lymphocytes to divide. The B- and T-cells have various surface glycoproteins. The plant mitogens are specific for the sugars present. PHA and Concanavalin-A stimulate T-cells. The same immobilized on Sephadex stimulate B-cells. Pokeweed mitogen induces blast transformation in both B- and T-cells.

Lymphoblasts proliferate and eventually differentiate into **effector cells** or **memory cells.**

Effector cells function in various ways to eliminate the antigen, and have short life spans. Memory cells look like small lymphocytes and can be distinguished from naive cells by the presence or absence of certain surface molecules. Plasma cells are the antibody secreting effector cells of the B-cell lineage.

Non-T and non-B lymphocytes (null cells): A majority of null cells possess F_cRs for IgG. These are known as the killer (K) cells and can exhibit ADCC (Antibody-Dependent Cell-mediated Cytotoxicity). This is a cytotoxic process mediated by effector cells linked to target cells by cell-membrane-bound antibody. The Natural Killer (NK) cells are large granular lymphocytes that display cytotoxic activity against a wide range of tumour cells and against cells infected with some viruses. They constitute about 5 per cent to 10 per cent of lymphocytes and have the ability to recognize tumour or virus-infected cells despite lacking antigen-specific receptors. NK-cells recognize potential target cells in two different ways. An NK-cell employs NK-cell receptors to distinguish abnormalities such as a reduction in the display of Class I MHC molecules, or it can distinguish the unusual profile of surface antigens shown by tumour cells or virus-infected cells.

These cells are heterogeneous with respect to surface markers. Receptors for F_c of IgG, for interferon and interleukin-2 are present. However, there are no receptors for foreign antigens nor are these cells MHC restricted. The natural killer cells do have certain unique receptors that are detected by monoclonal antibodies. NK-cells express a membrane receptor (CD16) for a specific region of the antibody molecule. They can therefore, attach to these antibodies and subsequently destroy the targeted cells. This process is known as Antibody-Dependent Cell-mediated Cytotoxicity (ADCC). They have the capability of being spontaneously cytotoxic to a variety of targets. Natural killing is enhanced in the presence of interferon (IFN). The NK-cells lyse the target cells more effectively in the presence of IL-2.

There is another cell type, the NKT-cell, which has some of the characteristics of both T-cells and NK-cells. NKT-cells have T-Cells Receptors (TCR), which interact with MHC like molecule called CD1 rather than with Class I or Class II MHC molecules. They have variable levels of CD16 and other receptors typical of NK-cells, and they can kill target cells. Activated NKT-cells can rapidly secrete large commands of cytokines needed to support antibody production by B-cells as well as information and the development and expansion of cytotoxic T-cells.

Antigen presenting cells

Foreign antigens must be specially processed before they can be presented to the antigen sensitive cells of the immune system. The most important antigen presenting cells are dendritic cells, macrophages and the B-cells (Figure 1.15). Antigen fragments generated inside these cells are bound to specialized receptors called MHC molecules in the cell cytoplasm and then transported to the surface of the antigen presenting cells.

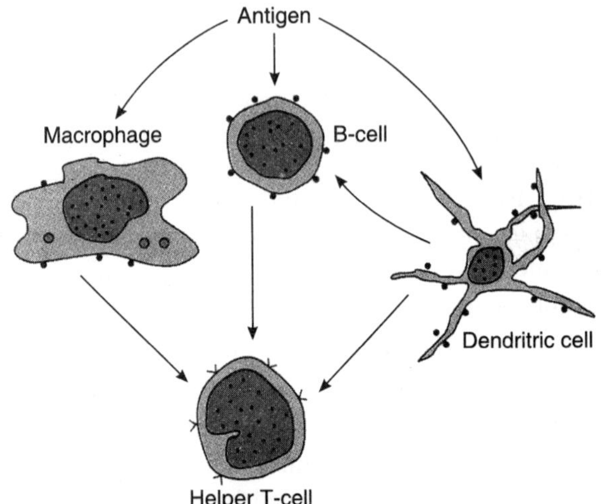

Figure 1.15 The three major groups of antigen-presenting cells.

The antigen presenting cells' main function is phagocytosis and cytotoxicity. They have the key ability to present a foreign antigen, in association with self-MHC antigen, to the antigen reactive lymphocytes. The APCs are needed in both humoral and cellular immunity for occurring efficiently.

Dendritic cells: These cells occur in many forms and perform distinct functions of antigen capture in one region and antigen presentation in another. Dendritic cells are located throughout the body, but specially in the lymphoid tissues like the Langerhans' cells of the skin, the medulla of the thymus, the lymph nodes and the spleen. These cells are large, adherent and motile cells, having numerous pseudopodia. They express Class II self-MHC antigens. The cells guard the body for signs of invasion by foreign pathogens and capture them when they enter. The dendritic cells engulf the antigen(s) by phagocytosis, or internalize them by receptor-mediated endocytosis, or by pinocytosis. They then migrate to the lymph nodes where they present the antigen to the T-cells. Mature dendritic cells lose the capacity for phagocytosis and large-scale pinocytosis. There is an increase of Class II MHC molecules in them and an increased production of co stimulatory molecules that are essential for the activation of naive T-cells.

The dendritic cells migrate to the lymph nodes after binding antigen and present the antigen to the T-cells in the paracortex. They also form cellular interconnections in the cortical B-cells area. All dendritic cells carry complement and antibody receptors on their surface, but are poorly phagocytic.

Dendritic cells present antigen in two ways. In an unprimed animal, the dendritic cells provide a surface on which antigen can be presented, thus making it a passive process. In animals that have had a previous exposure to antigen and thus, have some antibodies, there is a formation of an immune antigen-antibody complex. These immune complexes are taken up by the dendritic cells in the B-cell areas on their surface, and then are shed in round, beaded structures, called icosomes, which get subsequently attached to the B-cells.

Dendritic cells can retain antigen on their surface for more than three months, and antigens processed by dendritic cells are very potent stimulants of T-cells.

Macrophages: These are mononuclear phagocytes. There are monocytes circulating in the blood and macrophages in the tissues. The granulocyte-monocyte progenitor cells differentiate into pro-monocytes that leave the bone marrow and enter the blood, where they further differentiate into mature monocytes. Monocytes circulate in the blood stream for 8 hours during which they enlarge and then migrate into the tissues and differentiate into specific tissue macrophages.

The macrophages may also be formed by local proliferation in the tissues. They are larger than lymphocytes, with a diameter of 15 to 30 μm and possess a single, round nucleus and extensive cytoplasm (Figure 1.16). These cells are actively motile and are found throughout the body. Macrophages are intensely phagocytic and can phagocytose particles repeatedly. They are secretory cells and can secrete enzymes and soluble products that act on other cells. They produce proteins that influence inflammation, healing, and the body's response to infection. The most important of these proteins are the cytokines that include IL-1, IL-6, IL-12 and the tissue necrosis factor-α. IL-1 and IL-12 promote immune responses by enhancing lymphocyte responses to antigen; IL-6 influences the body's overall response to infection; and TNF-α destroys cancer cells.

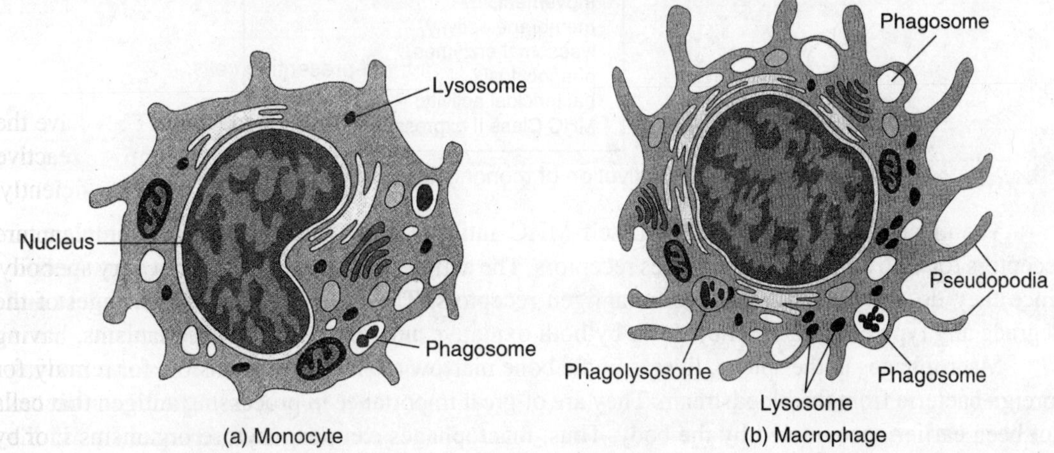

Figure 1.16 Typical morphology of a monocyte and a macrophage.

Unstimulated monocytes or macrophages are quiescent cells, which become rapidly activated on appropriate stimulation. Macrophages are activated by a variety of stimuli that includes cytokines secreted by activated T_h cells and by mediators of inflammatory response. An activated macrophage exhibits greater phagocytic activity, an increased ability to kill ingested microbes, and secrete more inflammatory mediators. Activated macrophages secrete various cytotoxic proteins that help them to eliminate a broad range of targets. They also express higher levels of Class II MHC molecules, and thus they function more effectively as antigen presenting cells for T_h cells. The macrophages engulf the foreign material with the help of pseudopodia. Fusion of the pseudopodia encloses the material within a membrane-bound structure called a **phagosome**, which then enters the endocytic processing pathway. The phagosome moves towards the cell interior where it fuses with a lysosome to form a **phagolysosome**. The hydrolytic enzymes in the lysosomes digest the ingested material of the phagolysosome and the digested contents are eliminated by exocytosis.

When the macrophages first move into inflamed tissues, they develop increased levels of lysosomal enzymes, enhanced phagocytic activity, higher expression of antibody and complement and higher secretion of neutral proteases. These inflammatory macrophages become further stimulated on exposure to bacterial products and proteins called **interferons**. The resulting activated macrophages have an enhanced ability to kill bacteria and some tumour cells by releasing reactive nitrogen metabolites (Figure 1.17).

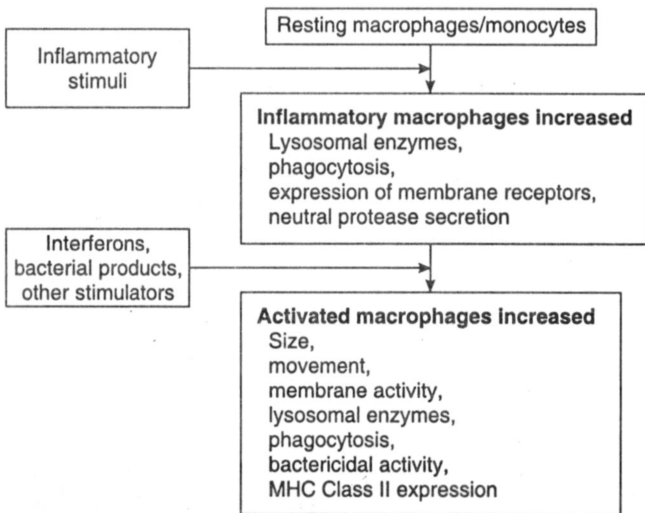

Figure 1.17 The activation of mononuclear phagocytic cells.

The macrophages possess Class II self-MHC antigens, F_cRs for IgG and IgE, complement receptors for C3b and also lymphokines receptors. The activity of macrophages is not very specific since they do not possess any specific antigen receptors. They can therefore, bind, ingest and degrade any type of antigen. They do so by both oxidative and non-oxidative mechanisms.

Macrophages in the spleen, liver, and the bone marrow are largely responsible for removing foreign bacteria from the bloodstream. They are of great importance in processing antigen that has not been earlier encountered by the body. Thus, macrophages can phagocytose organisms in the absence of antibody for opsonization.

Macrophages are necessary for the proper healing of wounds. When they arrive at a wound, they phagocytose and kill bacteria through secretion of lysosomal enzymes and oxygen metabolites. They also secrete elastose and collagenase that break down connective tissue. After the damaged tissue is removed, the macrophages help in the remodeling of the tissue by acting as a source of growth factors for fibroblasts and stimulate these fibroblasts to secrete collagen. Molecules that promote the growth of new blood vessels are also secreted by macrophages.

Other cells of the immune responses—the myeloid cells

The cells of the myeloid system are derived from the bone marrow ("myelos" is Greek for bone marrow). All the cells of this system have cytoplasm filled with granules, so they are collectively called **granulocytes**. These are the cells of the myeloid system whose cytoplasm is filled with granules.

Neutrophils: Sixty per cent of the circulating leukocytes are neutrophils. They are derived from pluripotent stem cells. Neutrophils are formed in the bone marrow; migrate to the blood stream, and about 12 hours later move into the tissues, where they live only for a few days.

Neutrophils are round cells about 12 μm in diameter (Figure 1.18) and contain a granulated cytoplasm that stains with both acid and basic dyes. The mature neutrophils have a 5-lobed nucleus, and are thus termed PolyMorphonuclear Neutrophils (PMNs). Neutrophils contain two types of cytoplasmic granules. The primary granules contain bactericidal enzymes such as peroxidase and lysozyme; neutral proteases like elastase; and acid hydrolases such as β-glucouronidase and cathepsin B. The secondary granules, which are smaller in size, contain enzymes such as lysozyme and collagenase, and the iron binding protein lactoferrin (Figure 1.19). The primary and secondary granules fuse with phagosomes, whose contents are ingested and eliminated. The proteins necessary for the function of neutrophils are formed during their development in the bone marrow. Their cytoplasm contains large amounts of glycogen, which is used for anaerobic glycolysis. Therefore, neutrophils can remain functional in damaged tissues where oxygen tension is low, such as sites of bacterial invasion. A variety of receptors (for Fc, C3b and C5a) are seen on the cell surface increasingly following the activation of these cells. The neutrophils carry out the phagocytosis of foreign, aberrant or dead cells. They also carry out pinocytosis of pathological immune complexes, as well as exhibit ADCC.

A number of substances generated in an inflammatory reaction serve as chemotactic factors for neutrophils at an inflammatory site.

Neutrophils have only a limited reserve of energy in the form of un-replenishable glycogen. They are therefore active immediately after being released from the bone marrow and can digest only a limited number of bacterial cells before they die through apoptosis. The dying neutrophil is destroyed by macrophages before they can release their lysosomal enzymes and cause tissue damage. Macrophages recognize the dying neutrophils on the basis of changes in the cell membrane lipids.

Neutrophils employ both oxygen-dependent and oxygen-independent pathways to generate antimicrobial substances.

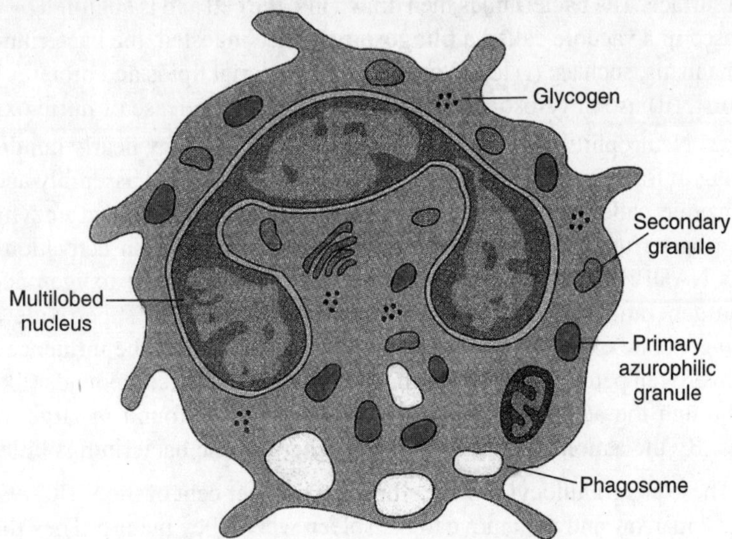

Figure 1.18 Typical morphology of a neutrophil.

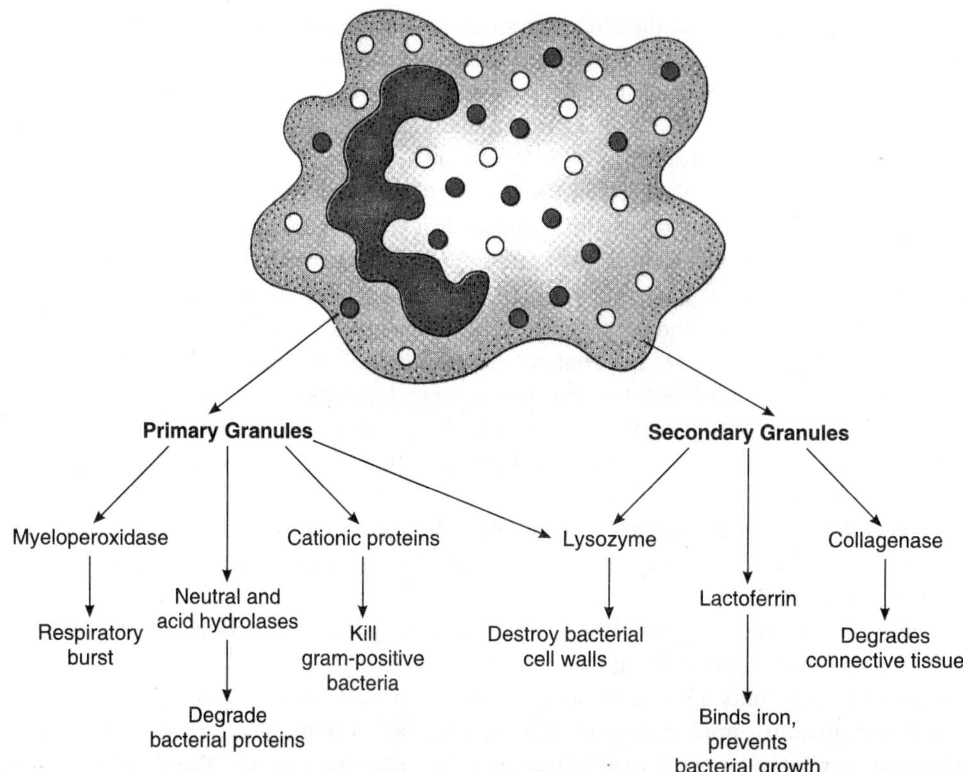

Figure 1.19 The major enzymes found within the primary and secondary granules of neutrophils and their functions in microbial killing and destruction.

When the pseudopodia of a neutrophil encounter a bacterium, it binds the bacterium firmly to the neutrophil surface. The bacterium is then drawn into the cell and is engulfed by the cytoplasm, where it is enclosed in a vacuole called a **phagosome**. Once ingested, the bacterium is exposed to destructive mechanisms, such as: (i) lethal oxidation of bacterial lipids and proteins by the process of respiratory burst, (ii) use of lysosomal enzymes, and (iii) the release of nitric oxide.

Respiratory burst. Neutrophils increase their oxygen consumption by nearly hundred fold within seconds of binding to foreign article. This increase is due to the rapid assembly and activation of a cell surface enzyme called NADPH-oxidase. This is a multi-component enzyme, which gets activated when a bacterium binds to the neutrophil Fc receptor. Upon activation the NADPH-oxidase converts NADPH to NADP+ in the presence of oxygen. The oxygen accepts a single donated electron generating a superoxide anion. Two molecules of superoxide anion interact spontaneously to generate one molecule of hydrogen peroxide under the influence of superoxide dismutase. The hydrogen peroxide is converted to other bactericidal compounds (like chloride and hypochlorate) through the action of myeloperoxidase, which is found in large amounts in the primary granules. By the action of the free radicals generated the bacterium is killed.

Eosinophils: These are granulocytes that comprise 3 to 5 per cent of the WBCs. Eosinophils are formed in the bone marrow and then move to the spleen where they mature. They then circulate in the blood, where their half-life is 30 minutes, before they migrate into the tissues where they have a half-life of about 12 days. The cells have bilobed nuclei, abundant ribosomes and mitochondria (Figure 1.20).

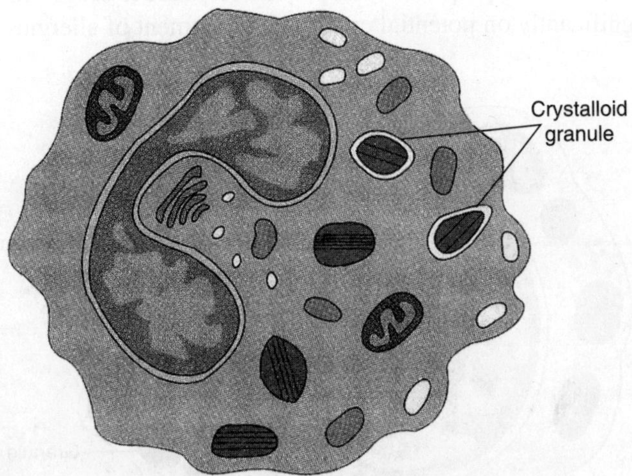

Figure 1.20 Typical morphology of an eosinophil.

Eosinophils are motile phagocytic cells. They defend against parasitic organisms by secreting the contents of eosinophils granules that damage the parasite membrane.

Eosinophils participate in phagocytosis, pinocytosis and ADCC through FcRs. They can ingest and destroy foreign material. The eosinophils contain large quantities of acid-phosphatase and peroxidase, and respond to chemotactic factors released by T-cells, basophils and mast cells. These eosinophils can degranulate and release toxic proteins, as well as release enzymes that inactivate mediators of Type I hypersensitivity. Each large eosinophil granule has a highly toxic protein called Major Basic Protein (MBP), which destroys invading helminthes. Like neutrophils, eosinophils can destroy invaders through respiratory burst.

Basophils and mast cells: **Basophils** have a major role to play in certain allergic responses. These cells make up less than 1% of the WBC. They contain basophilic granules, which contain eosinophil chemotactic factors and mediators of Type I hypersensitivity. Upon entry into an organism, the allergens bind to IgE and cross-link them. They then attach to the basophils through FcRs for IgE, making the granules release their contents. The cells also contain FcRs for IgG and receptors for C3a, C3b and C5a, due to which they have the capability to exhibit ADCC, phagocytosis and chemotaxis. Basophils infiltrate tissues under the influence of lymphocytes, where they provoke inflammation since their granules contain vasoactive amines like histamine and serotonin.

Mast cells are found in various tissues including the skin, collective tissue of various organs, and mucosal epithelial tissue of the respiratory, genitourinary and digestive tracts. They have a large number of granules if that contain histamine and other active substances. Mast cells too function in the development of allergy.

Mast cells are similar to the basophils in their morphology, receptors and functions (Figure 1.21). The mast cells are not homogenous, having two sub-populations differing critically in their physiology and morphology. One type of mast cell is present in the mucosa of the lungs and intestines, while the other type is found in the connective tissues near blood vessels. These two sub-populations also differ in the number and size of granules, density of receptors expressed on the cell surface, pharmacological properties, lifespan and response to drugs. How these cells respond to drugs can bear significantly on potential control and treatment of allergies.

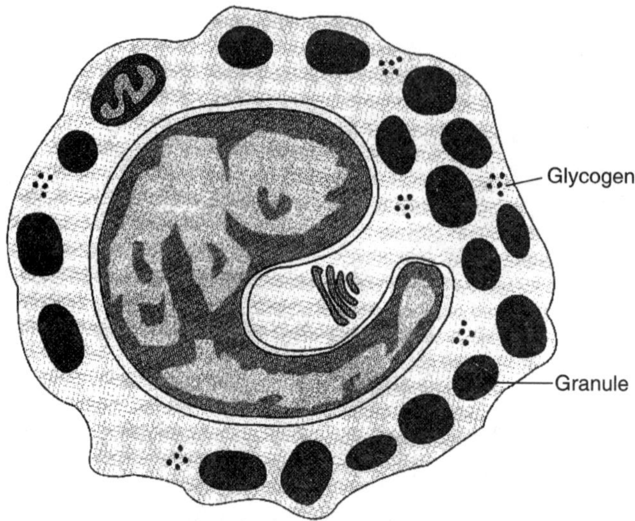

Figure 1.21 Typical morphology of a basophil.

Platelets: Platelets are small cytoplasmic fragments, having a diameter that is half of RBCs. They possess small granules, FcRs for IgG and IgE as well as receptors for Class I MHC antigens.

Platelets have a very important role in blood clotting, hypersensitivity reactions and inflammatory process. These cells can release permeability increasing substances like histamine. When activated by platelet activating factor, they can aggregate and degranulate. Blood platelets can adhere to and engulf some bacteria.

MULTIPLE-CHOICE QUESTIONS

1. The basic function of cellular immunity is to:
   (a) fight virus infections
   (b) reject grafts
   (c) destroy abnormal cells
   (d) fight intracellular bacteria
2. Opsonin is a:
   (a) granulocyte
   (b) chemokine
   (c) lysosomal enzyme
   (d) substance that enhances phagocytosis

3. The cell surface molecule that begins the respiratory burst is:
   (a) myeloperoxidase
   (b) singlet oxygen
   (c) NADPH-oxidase
   (d) superoxide dismutase
4. When neutrophils die they do so by a process called:
   (a) phagocytosis
   (b) necrosis
   (c) endocytosis
   (d) apoptosis
5. In the spleen, the B-lymphocytes are largely found in the:
   (a) follicular areas
   (b) periartiolar lymph sheath
   (c) red pulp
   (d) marginal zone
6. In the lymph node the T-cells are mainly found in:
   (a) primary follicles
   (b) paracortex
   (c) sinusoids
   (d) germinal centre
7. In the lymph nodes, the memory cells are generated in:
   (a) primary follicles
   (b) paracortex
   (c) sinusoids
   (d) germinal centre

REVIEW QUESTIONS

1.1 List the most important enzymes found inside the neutrophils.

1.2 What would be the effect of removal of the spleen, a lymph node and the thymus from a newborn mouse?

1.3 How do the circulating patterns of T- and B-lymphocytes reflects their functions? Is there any advantage of having circulating T-lymphocytes?

1.4 Write notes on the following:
   (i) B-lymphocytes
   (ii) T-lymphocytes
   (iii) Macrophages
   (iv) Dendritic cells
   (v) Natural killer cells

1.5 Give an account of primary lymphoid organs.

1.6 What are secondary lymphoid organs? How do they differ from the primary lymphoid organs?

2

Antigens

2.1 ANTIGENS: AN OVERVIEW

The function of the immune system is to protect the body against foreign invaders. It is, therefore, imperative that cells must have the mechanisms to recognize these invaders.

Human beings are exposed to a great variety of foreign material in everyday life and not all of it is threatening. We consume lots of protein, carbohydrate and fat as food. These are generally not a threat to the body and are not usually regarded as foreign. Similarly, large inert foreign bodies such as metal bone pins or plastic heart valves fail to provoke an immune response. There are some essential features required of a foreign molecule in order for it to be recognized by the cells of the immune system, act as an antigen, and provoke an immune response.

There are two major restrictions on antigenic molecules. First the molecules must be recognized as foreign. Second, because of the processing that antigens must undergo, there are physical and chemical limitations on the types of foreign molecules that can stimulate immune response. The most effective antigens are large, rigid, chemically complex molecules that can be degraded fairly readily to soluble fragments, which the immune system can recognize.

A foreign molecule that is specifically recognized by the host's immune system as foreign or non-self and is capable of triggering an immune response is called an immunogen. An antigen can specifically bind to an antibody molecule. An antigen that is capable of producing an immune response is referred to as an **immunogen**. The term **antigen** is used for a molecule that is only capable of binding to an antibody, but that does not necessarily induce an immune response. Therefore, all immunogens are considered antigens, but not all antigens are immunogens.

Nearly all biological molecules like sugars, lipids, hormones and macromolecules can function as an antigen or possess the potential for being antigenic. The portion of an antigen that is recognized by the T- and B-lymphocytes and also binds to the specific antibodies are known as an **antigenic determinant** or an **epitope**. An epitope is the smallest biochemical unit of an antigen that is capable of eliciting an immune response. It's size and structure varies according to the type of the antigen.

B-cells + antigen → effector B-cells + memory B-cells
↓
(plasma cells)

T-cells + antigen → effector T-cells + memory T-cells
↓
(e.g. CTLs, T_h-cells)

The number of different antibodies which may be produced to an antigen is high, because the antigens are three-dimensional structures and present many different configurations to the B-cells. Different antibodies to an antigen often bind to epitopes, which overlap on the antigen surface. In this way different antibodies can bind to a particular antigenic region of the molecule, without binding to exactly the same epitope.

The protein content of certain bacteria and most viruses induce strong humoral and cell-mediated responses, and produce a long-lasting immunologic memory. Recognition of protein antigens by T-lymphocytes is MHC restricted.

Polysaccharides and lipids are not recognized by T-cells and therefore, are not capable of inducing cell-mediated immune responses by stimulating MHC-restricted T-cells. However, an induction of T-cell-independent humoral immune response may occur, with a production of IgM antibodies that provide only a short-lived immunity.

2.2 REQUIREMENTS FOR IMMUNOGENICITY

For a molecule to be immunogenic, certain criteria have been defined they include the following:

Foreignness

A molecule must be recognized as non-self by the biological system to elicit an immune response. Antigens that have not been presented to immature lymphocytes during their development are later recognized as non-self or foreign by the immune system. The degree of immunogenicity depends on the degree of its foreignness. The greater the phylogenetic distance between two species, the greater the structural, and therefore, the antigenic disparity between their constituent molecules. Thus, bovine serum albumin is not immunogenic when injected into a cow, but is strongly immunogenic when injected into a rabbit. There are some exceptions to this rule. Macromolecules like collagen and cytochrome c have been highly conserved throughout evolution and therefore, display very little immunogenicity across species. On the other hand some self-components, like corneal tissue and sperm, are effectively sequestered from the immune system, so that if these are injected into the animal from which they originated they will function as immunogens. The development of antisperm antibodies is a common sequel to vasectomy as a result of leakage of sperm antigens into the tissues.

The ability of the immune system to specifically recognize and respond to a foreign molecule depends on the host's inheritance of a specific set of major histocompatibility genes. The MHC genes code for Class I and Class II MHC molecules that are expressed on the surface of all nucleated and immunocompetent cells.

Accessibility

The epitopes must be readily accessible to the immune system for recognition. Although it is possible to produce antibodies to almost any part of an antigen, this does not normally happen in an immune response. It is usually found that certain areas of the antigen are particularly antigenic, and that majority of the antibodies bind to these regions. These regions are the immunodominant regions of the antigens, and are at exposed areas on the outside of the antigen. The parts of the peptide chains, which protrude significantly from the globular surface, tend to be sites of high epitope density. Hence, accessibility of these high epitope density areas to the recognition system determines the outcome of the immune system. The immunodominant regions of the antigen also correspond to the most mobile surface areas of the molecule.

Size of the molecule

Macromolecules have the ability to initiate a humoral immune response. These macromolecules are larger than the antigen-binding region on an antibody molecule and possess multiple epitopes, each of which may be bound to an antibody. The epitopes or antigenic determinants are molecular sites that stimulate immune responses. The generally accepted minimal size of a molecule that is able to induce an immune response is 10,000 daltons on the basis of molecular weight. In general, large molecules are better antigens than small molecules. The most active immunogens have a molecular mass of greater than 100,000 daltons. Substances with a molecular mass < 5000 to 10,000 daltons are poor immunogens, although a few substances with a molecule mass < 1000 daltons have seen to be immunogenic. Para-azobenzene-arsenate trityrosine (750 daltons) is the smallest antigen which is used to provoke antibodies in guinea pigs and rabbits.

Structure of the molecule

The general rule is that the more complex a molecule, the greater is its immunogenicity. Small peptides or degraded proteins must first be bound to a MHC molecule for recognition and response by the immune system. These small peptides cannot be recognized in the free state. Synthetic homopolymers, which are composed of multiple copies of a single amino acid or sugar, lack immunogenicity regardless of their size. Heteropolymers are usually more immunogenic. Hence, chemical complexity contributes to immunogenicity. All four levels of protein organization contribute to structural complexity of protein and therefore, affect its immunogenicity. Lipids can serve as haptens attached to suitable carrier molecules like bovine serum albumin.

The cells of immune system must recognize the specific shape of a molecule to consider it foreign. Therefore, highly flexible molecules that have no fixed shape are poor antigens. An example is gelatin, which is a protein with structural instability; this is a poor antigen until it is stabilized by the incorporation of tyrosine or tryptophan molecules which cross-link the peptide chains. Similarly, flagellin, which is a major protein of bacterial flagella, is a structurally unstable weak antigen. Starch and other simple repetitive polysaccharides are poor antigens because they do not assume a stable configuration. For the very same reason proteins are better antigens than large, repeating polymers like the lipids, carbohydrates, and nucleic acids.

Antigen dose

The optimal dose for immunogenicity depends on the type of the antigen. Very small or large doses of an antigen may produce a state of unresponsiveness, known as immunologic tolerance. Some combination of optimal dosage and route of administration will induce a peak immune response in a given animal. A single dose of most experimental immunogens does not induce a strong response; instead a repeated administration over a period of weeks is required to stimulate a strong immune response.

Degradability

Large, insoluble, or aggregated molecules generally are more immunogenic than small, soluble ones because the larger molecules are more readily phagocytosed and processed. The cells of the immune system, however, recognize small molecule fragments and soluble antigens. If a molecule cannot be broken up or solubilized then it cannot act as an antigen. Macromolecules that cannot be degraded and presented with MHC molecules are poor immunogens. Polymers of D-amino acids cannot be processed and therefore are poor immunogens. If a few short peptides containing L-amino acids are inserted into such polymers, the resulting molecule is a good immunogen. Stainless steel pins and plastic joints are commonly implanted in bodies without triggering an immune response. The lack of antigenicity is due to their inertness; they cannot be fragmented and processed to a suitable form for triggering an immune response.

2.3 CONTRIBUTION OF BIOLOGICAL SYSTEM TO IMMUNOGENICITY

The genotype of the recipient is a major factor in determining immune responsiveness. Hugh McDevitt showed that two different inbred strains of mice responded very differently after exposure to a synthetic polypeptide immunogen. They showed different levels of serum antibody. When the two strains were crossed, the F_1 generation showed an intermediate response to the immunogen. Genetic control of immune responsiveness is confined to genes within the MHC. The MHC proteins function in the presentation of processed antigen to T-cells. They play a central role in determining the extent to which an animal responds to an immunogen. The genes that encode B-cell and T-cell receptors also influence the response of an animal to an antigen. Thus, genetic variability of the genes affects the immunogenicity of a given macromolecule.

Some combination of optimal dose and route of administration induces a peak immune responses in a given animal. An insufficient dose fails to activate enough lymphocytes and induces tolerance. An excessively high dose can also induce tolerance. The administration route strongly influences the immune organs and cells that will be involved in the response. Antigens that are administered intravenously are carried first to the spleen, while antigens that are administered subcutaneously goes first to the local lymph nodes.

2.4 ADJUVANTS

These are substances which, when mixed with an antigen and injected with it, non-specifically enhance the immunogenicity of that antigen. Adjuvants are often used to boost the immune response when the antigen has low immunogenicity. Some of the known adjuvants are now recognized as ligands for the toll-like receptors on the dendritic cells and macrophages and thus stimulate immune responses through the activation of the innate immune system. Adjuvants exert one or more of the following effects:

(a) Prolong antigen persistence
(b) Enhance co-stimulatory signals
(c) Induce granuloma formation
(d) Stimulate non-specific proliferation of lymphocytes.

Aluminium hydroxide, alum (aluminium potassium sulphate) and Freund's adjuvant are the commonly used adjuvants to enhance the immunogenicity of antigens. The most potent adjuvant is Complete Freund's Adjuvant (CFA). Aluminium potassium sulphate (alum) is the only adjuvant approved for general human use. Alum precipitates the antigen, resulting in slower release of antigen from the injection site; it also increases the size of antigen thereby increasing the likelihood of phagocytosis. Water-in-oil adjuvants (known as incomplete Freund's adjuvant) prolongs the persistence of antigen. It contains an antigen in aqueous solution, mineral oil, and an emulsifying agent like mannide monooleate, which disperses the oil into small droplets surrounding the antigen. Complete Freund's adjuvant contains heat-killed mycobacteria as an additional ingredient. Muramyl dipeptide, which is a component of the mycobacterial cell wall, activates dendritic cells and macrophages making complete Freund's adjuvant far more potent than the incomplete form. Antigen presentation and the requisite costimulatory signal are usually increased in the presence of adjuvant.

2.5 EPITOPES

The part of the antibody molecule, which makes contact with the antigen, is termed the **paratope**. Consequently, the part of the antigen molecule that makes contact with the paratope is called the **epitope.** As most antigens are protein in nature, they exist in a folded three-dimensional, tertiary structure. Hence there may be a cluster of amino acid sequences on the three-dimensional structure constituting a series of epitopes. Each of these epitope clusters is also known as **antigenic determinant**.

Epitopes are the immunologically active regions of an immunogen that bind to antigen-specific membrane receptors on lymphocytes or to secreted antibodies. Most antigens are structurally complex, containing different epitopes, and the immune system usually responds by producing antibodies to several epitopes on the antigen. Therefore, several B-cell clones are stimulated and proliferate.

The immune cells do not interact with or recognize the entire immunogen molecule. The lymphocytes recognize only discrete sites on the macromolecule called the epitopes or antigenic determinants. B- and T-cells recognize different epitopes on the same antigenic molecule. When mice are immunized with glucagon, which contains 29 amino acids, antibody is elicited to the epitopes in the amino terminal portion, whereas the T-cells respond only to the epitopes in the carboxyl terminal portion. In polysaccharides, the branch points may contribute to the conformation of epitopes. B-cells recognize soluble antigen when it binds to antibody molecules on the B-cell membrane. Antigen derived from enzymatic digestion of pathogen proteins are recognized by the T-cell receptors only when it is complexed with an MHC antigen.

Antigens recognized by T-cells must therefore, have two distinct interaction sites: one is the epitope that interacts with the T-cell receptor, and the other called the **agretope** interacts with an

MHC molecule. Immunodominant T-cell epitopes are determined in part by a set of MHC molecules expressed by an individual.

When interacting with antigens, B-cells involve binary complex of membrane Ig and antigen. They can bind a soluble antigen. B-cells react to protein, polysaccharides and lipids, and do not require the involvement of MHC molecules. To be recognized by B-cells the epitopes should be assessable, hydrophilic, mobile peptides containing sequential or non-sequential amino acids.

Antigen recognition by T-cells on the other hand, involve ternary complex of T-cell receptor, antigen and MHC molecule. The involvement of MHC molecules is required only by T-cells to display processed antigen. T-cells consider mostly proteins as antigens, but some lipids and glycolipids presented on MHC like molecules are also considered as antigens by T-cells. The epitopes are usually internal linear peptides produced by processing of antigen and bound to MHC molecules.

B-cells recognize soluble antigen when it binds to their membrane bound antibody. The B-cell epitopes on native proteins are generally composed of hydrophilic amino acids on the protein surface that are topographically accessible to membrane-bound or free antibody. These epitopes can contain sequential or non-sequential amino acids. The B-cell epitopes tend to be located in flexible regions of immunogen and display site mobility. Complex proteins contain multiple overlapping B-cell epitopes, some of which are immunodominant.

T-cells recognize only peptides combined with MHC molecules on the surface of antigen-presenting cells and altered self-cells. T-cell receptors do not bind free peptides. Antigenic peptides recognized by T-cells form termolecular complexes with a T-cell receptor and an MHC molecule.

2.6 HAPTENS

Small organic molecules that are antigenic, but not immunogenic are called **haptens**. Chemical coupling of a hapten to a large protein, called a **carrier**, yields an immunogenic **hapten-carrier conjugate**. The resulting hapten-carrier complex serves as an immunogen for initiating a humoral immune response. Antisera made against such a conjugate has antibodies specific for (i) the hapten determinant, (ii) unaltered epitopes on the carrier protein, and (iii) new epitopes formed by combined parts of both the carrier and the hapten in combination.

When multiple molecules of a single hapten are coupled to a carrier protein, the hapten becomes accessible to the immune system and often functions as the immunodominant determinant. Many biologically important substances, including drugs, peptide hormones, and steroid hormones, can function as haptens. The hapten specific antibodies are useful for measuring the presence of various non-immunogenic substances in the body. The study of haptens has allowed us to learn about the structural basis of antibody specificity.

Haptens are, therefore, small well-defined chemical groups such as dinitrophenol (DNP), which are not immunogenic on their own, but will react with preformed antibodies. To make a hapten immunogenic, it must be linked to a carrier molecule, which is immunogenic.

The hapten cannot function as an immunogenic epitope by itself. When multiple molecules of a single hapten are coupled to carry protein the hapten is recognized by the immune system and can then function as an immunogen. Antibodies produced against these haptens are useful for measuring the presence of various substances in the body.

2.7 SUPERANTIGENS

Superantigens are bacterial or viral proteins that can bind simultaneously to Vβ region of TCR and the α chain of Class II MHC molecules. Superantigens can be exogenous or endogenous, both of which are capable of cross-linking TCR and Class II MHC molecules, and provide a signal for activation and proliferation of T-cells. Exogenous antigens are soluble antigens secreted by bacteria. A variety of exotoxins of gram positive bacteria are exogenous antigens. Endogenous superantigens are cell membrane proteins encoded by many viruses. Certain superantigens can bring about T-cell activation and also maturation of T-cells in the thymus. These superantigens bind to TCR Vβ of the thymocytes and when these thymocytes interact with thymic stromal cells, they lead to deletion of these thymocytes.

MULTIPLE-CHOICE QUESTIONS

1. An epitope is about the size of:
(a) one amino acid residue
(b) three amino acid residues
(c) five amino acid residues
(d) ten amino acid residues

2. A hapten is a(n):
(a) a small molecule attached to a protein
(b) antibiotic
(c) amino acid side chain
(d) carbohydrate side chain

3. Plastics are not good antigens because they are:
(a) toxic
(b) synthetic
(c) degraded too slowly
(d) complexed with proteins

REVIEW QUESTIONS

2.1 List the properties of an ideal antigen.
2.2 What is the role of the biological system in immunogenicity?
2.3 What are adjuvants? Give their properties.
2.4 Define haptens, antigens and immunogens.
2.5 Write an account on superantigens.

3

IMMUNOGLOBULINS: STRUCTURAL AND FUNCTIONAL BASIS OF ANTIBODY DIVERSITY

3.1 IMMUNOGLOBULINS

Immunoglobulins are a group of glycoproteins present in the serum and tissue fluids of all mammals. The immunoglobulins found in the serum are the mixture of molecules with antibody activity against a wide variety of antigens and their epitopes. Some immunoglobulins are carried on the surface of B-cells, where they act as receptors for specific antigens. Other antibodies are free in the blood or lymph. Contact between B-cells and antigen is needed for the B-cells to develop into Antibody Forming Cells (AFCs) or plasma cells that secrete large amounts of antibody.

Plasma cells normally form only one specific kind of antibody in relatively small quantities. Sometimes plasma cells become cancerous, resulting in the growth of a clone of cancerous plasma cells that produce single, absolutely homogeneous **(monoclonal)** immunoglobulins. A plasma cell tumour is known as a **myeloma** and its monoclonal immunoglobulin product is called myeloma protein, which is secreted in large quantities into the bloodstream of affected individuals. Each monoclonal tumour has identical cells. Myeloma cells can be cultured to produce large quantities of pure monoclonal antibodies. If mature plasma cells are fused with myeloma tumour cells, hybridoma cells are obtained. These cells are immortal and they produce huge quantities of antibodies. They express the B-cell's immunoglobulins and the properties of the myeloma cells.

Immunoglobulins are glycoproteins synthesized by B-cells in two forms: (i) B-cell receptors that bind extracellular (non-processed) antigens and serve as B-cell markers and (ii) the secreted form, generally known as the antibodies.

Immunoglobulins and antibody molecules are terms that are often used interchangeably. Immunoglobulins refer to all forms of Ig molecules, that is, secreted antibodies and antigen receptors that are known as B-cell surface immunoglobulins. Antibodies refer to the secreted forms of Igs that are produced by the plasma cells (B-cell lineage) and secreted into body fluids, such as blood.

In membrane-bound immunoglobulins the carboxyl terminal domain differs in structure and function, from that of the secreted immunoglobulin. The carboxyl terminal domain of membrane immunoglobulin has a hydrophilic extracellular spacer sequence of 26 amino acid residues, a transmembrane hydrophobic sequence and a short tail directed into the cytoplasm.

The serum antibodies produced in response to a particular antigen are heterogeneous. Because most antigens are complex and contain many antigenic determinants, the immune system usually responds by producing antibodies to several of them. **Circulating antibodies** recognize antigen in serum and tissue fluids.

3.2 HISTORY

In the 1950s, experiments by Porter and Edelman revealed the basic structure of the immunoglobulin molecule. IgG was subjected to papain, pepsin and mercaptoethanol digestion and the various fractions produced were analyzed by separating them through column chromatography. The 150,000 daltons IgG molecule was found to be composed of two 50,000 daltons polypeptide chains designated as heavy (H) chains and two 25,000 daltons chains designated as light (L) chains. Each immunoglobulin molecule has got two Fab sites (that bind to antigen) and one Fc site that crystallizes under freezing conditions. Porter raised antiserum to these two portions and found that antibody to Fab fragment could react with both the H and L chains, whereas antibody to the Fc fragment reacted only with the heavy chain. From this it was concluded that Fab consists of portions of a heavy and light chain while the Fc contains only heavy chain component.

3.3 STRUCTURE OF IMMUNOGLOBULIN MOLECULES

The basic structure of all immunoglobulin molecules is a unit consisting of two identical light polypeptide chains and two identical heavy polypeptide chains, linked together by covalent (disulphide bonds) and non-covalent forces. The heavy chains differ between various classes of immunoglobulins (Figure 3.1). Two types of immunoglobulins, IgA and IgM, also occur as oligomers of the four-chain unit.

The light chain has a molecular weight of 25,000 and is common to all classes, whereas the heavy chain has a molecular weight of 55,000–77,000 and is structurally different for each class or subclass.

The light chains of vertebrates have two distinct forms called kappa (κ) and lambda (λ). The heavy chains of the vertebrates have five forms called γ, μ, α, δ and ε. Either of the light chain types may combine with any of the heavy chain types, but in any one immunoglobulin molecule both light chains are identical as are the two heavy chains. The light chain consists of two distinct regions. The C-terminal half of the chain, which has approximately 107 amino acid residues, is constant and is called the C_L (Constant: Light chain) region, whereas the N-terminal half of the chain shows sequence variability and is known as the V_L (Variable: Light chain) region. Thus, all light chains have one variable and one constant region.

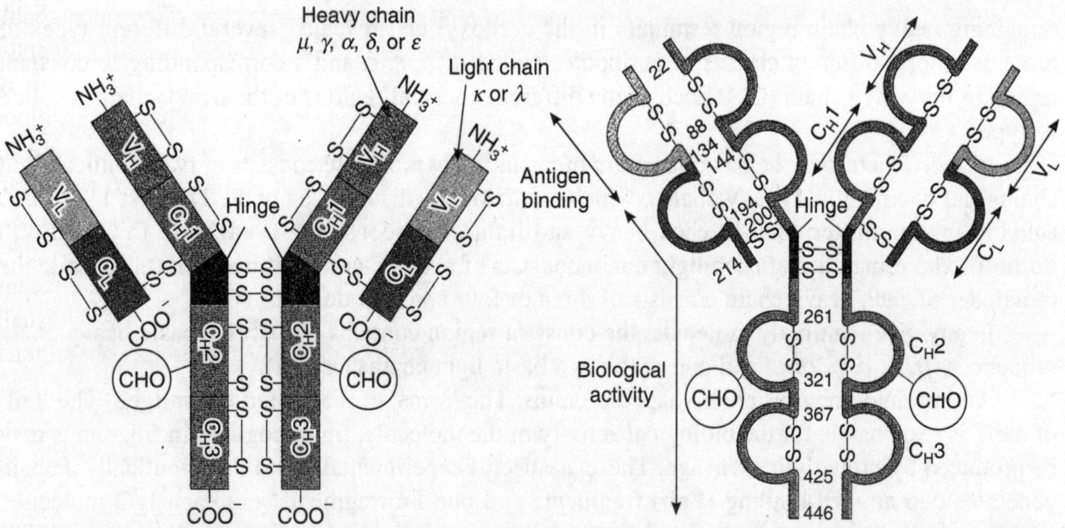

Figure 3.1 Schematic diagram of the structure of immunoglobulins.

A typical antibody molecule has two intrachain disulphide bonds in the light chain—one each in the variable and constant region. Heavy chains, which are twice the length of a light chain, have got four such bonds. Each disulphide bond encloses a peptide loop of 60–70 amino acid residues. Each peptide chain of the immunoglobulin has a series of globular regions with similar secondary and tertiary structure. The chains are folded into discrete regions called **domains.** Each of the domains in the immunoglobulin molecule has a characteristic tertiary structure called the **immunoglobulin fold.**

There are two domains in light chain and four or five in the heavy chains, depending upon their class. The peptide loops enclosed by the disulphide bonds represent the central part of a 'domain' of about 110 amino acid residues. The first of these domains correspond to the **variable region** of the heavy and light chains and are designated V_H and V_L respectively. The variable region is also the antigen-binding region. The heavy chains of IgG, IgA and IgD have three more domains that make up the **constant region** of the chain, C_H1, C_H2 and C_H3. The constant region is associated with biological activity or the effector functions attributed to each class of immunoglobulins. IgM and IgE have an extra constant domain in their heavy chain, thus having four constant domains each.

Light chain sequencing. The amino terminal half of the light chain consists of 100–110 amino acids and showed lots of variation among different immunoglobulins. This region is called the variable (V_L) region. The carboxyl terminal half of the light chain is called the constant (C_L) region and has two different types of protein chains among different immunoglobulins. They are the kappa and lambda chains. In humans, 60 per cent light chain constant regions are kappa and 40 per cent are lambda. The four IgG subtypes in humans differ from one another by interchange of two or three amino acid positions in the kappa or lambda chain. A single antibody molecule contains either kappa or lambda chains but never both.

Heavy chain sequencing. The amino terminals consisting of 100–110 amino acids show variation among different antibodies and are called as variable region of the heavy chain (V_H). The remaining heavy chain region terminate in the carboxyl end revealed several different types of proteins among different classes of antibodies called μ, δ, ε, α and γ corresponding to constant region of the heavy chain (C_H). Each of the different constant regions of the heavy chain is called isotype.

In short, therefore, the basic structure of an antibody molecule consists of two identical light chains and two identical heavy chains, which are linked by disulphide bonds. The first 110 amino acids at the amino terminal in each heavy and light chain constitute a **variable (V_H and V_L) domain**. The remainder of each light chain consists of a single **constant domain** (C_L), while the remainder of each heavy chain consists of three or four constant domains (C_H).

In any given antibody molecule, the constant region contains one of five basic heavy-chain sequences (μ, λ, δ, α, or ε) and one of the two basic light chains (κ or λ).

All immunoglobulins are Y-shaped proteins. The 'arms' of the Y bind the antigen. The 'tail' of the Y is responsible for the biological activity of the molecule. Immunoglobulin fragments may be produced by proteolytic cleavage. These are useful experimentally and therapeutically. Papain generates two antigen-binding **(Fab)** fragments and one **Fc** fragment from each IgG molecule, whereas pepsin produces a large **F (ab')** $_2$ fragment containing both antigen binding sites.

The ability of an immunoglobulin to bind is determined by the sequence of amino acids in its variable regions. This sequence determines the shape of the antigen-binding site or **paratope.** Within the variable regions are highly variable amino acid sequences making the **hypervariable region** (antigen binding site). Each hypervariable region contains three highly variable amino acid sections, known as the **Complementarity Determining Regions**, or **CDR1** (beginning at the N terminus), **CDR2**, and **CDR3** (showing most variability). As the variable regions fold into globular domains in a three-dimensional configuration, the CDRs form protruding loops on the surface of the domain that are complementary to the configuration of the antigen with which the immunoglobulin can bind. It is therefore, these parts that determine the specificity of an antibody for an antigen. The hypervariable regions form the antigen binding sites. There are three such regions in the V domains of each light and heavy chain. The folding of the domains causes them to be clustered at the distal tips of the molecule, producing two antigen-binding sites for each four-chain unit. These regions are located around amino acid positions 30, 50, and 95. The CDRs are flanked by relatively constant regions called Framework Regions (FRs). These determine the fold that ensures that the CDRs are in proximity to each other. The work of the FRs is to 'push or pull' the CDRs to accommodate the epitopes during binding of antigen to antibody.

The γ, δ and α heavy chains contain an extended peptide sequence between the first and second constant regions (C_H1 and C_H2) that has no homology with other domains. This is the **hinge region**, which is rich in proline residues. It provides the immunoglobulin molecules of type IgG, IgD and IgA with segmental flexibility necessary for binding with antigens. The angles between the arms of the Y-shaped antibody molecule differ in different complexes, reflecting the flexibility of the hinge region. Another prominent amino acid present in this region is cysteine, which forms interchain disulphide bonds that hold the two heavy chains together. IgM and IgE lack the hinge region; instead they have an additional domain, consisting of 110 amino acids, that has hinge like features.

A **J chain** is present in all polymeric forms of immunoglobulins. This is a small polypeptide, which is covalently bonded to the heavy chains of IgM (pentamer) and IgA (dimer). The presence of the J chain is responsible for the polymerization of the basic Ig units.

3.4 FUNCTIONS OF ANTIBODIES

All antibodies are bifunctional. They exhibit one or more **effector functions** in addition to antigen binding. These biological activities (e.g. complement activation and cell binding) are localized to sites that are distant from the antigen binding sites (mostly on the Fc region). The antibodies have the ability to:

1. Specifically bind with antigens through the variable region of the Ig molecule (a function exhibited also by the B-cell surface Igs).
2. Trigger various class-specific biological activities, a function unique to the constant region of the antibody molecules.

3.5 FUNCTIONAL REGIONS OF ANTIBODY MOLECULES

The Fab portion contains the variable region which binds antigen.

Fc portion mediates antigen clearance by binding to complement and to Fc receptors on the cells of the immune system.

Membrane spanning region is heavy chain carboxyl terminal domain, present only in immunoglobulins expressed on the surface of B-cells.

The recognition of foreign antigen is the hallmark of the specific adaptive immune response. Two distinct type of molecules are involved in this process—the immunoglobulins and the B-cell antigen receptors.

3.6 CLASSES OF IMMUNOGLOBULINS

Immunoglobulins are a family of glycoproteins. In humans, five distinct classes of immunoglobulin molecules are seen namely IgG, IgM, IgA, IgD and IgE (Figure 3.2) which differ in sizes, charge, amino acid sequence and carbohydrate content. The IgG class is divided into four subclasses (IgG1, IgG2, IgG3 and IgG4) while the IgA has two subclasses (IgA1 and IgA2). The classes and subclasses represent nine isotypes which are seen in all normal individuals. Each isotype is defined by the sequence of the constant region of its heavy chain. There are no subclasses of IgM, IgD or IgE.

Immunoglobulin class and subclass depends on the structure of heavy chain. It is the heavy chain type that determines the class and subclass of an immunoglobulin molecule. The immunoglobulins within each class are also very heterogeneous. These differences are genetically determined.

3.6.1 Immunoglobulin G

This is the major immunoglobulin in the normal human serum, accounting for 70 –75 per cent of the total immunoglobulin pool. This has a molecular weight of about 150,000 and a sedimentation coefficient of 7S. The half-life of IgG in the body is 18 to 23 days and the carbohydrate content is 3 per cent. IgG is the major antibody of secondary immune responses and the exclusive antitoxin class.

Maternal IgG also confers immunity in neonates. IgG has four subclasses, IgG1 through IgG4, which are distinguished from one another by the interchain disulphide linkages between the heavy chains and the length of the hinge region. There are also subtle amino acid differences

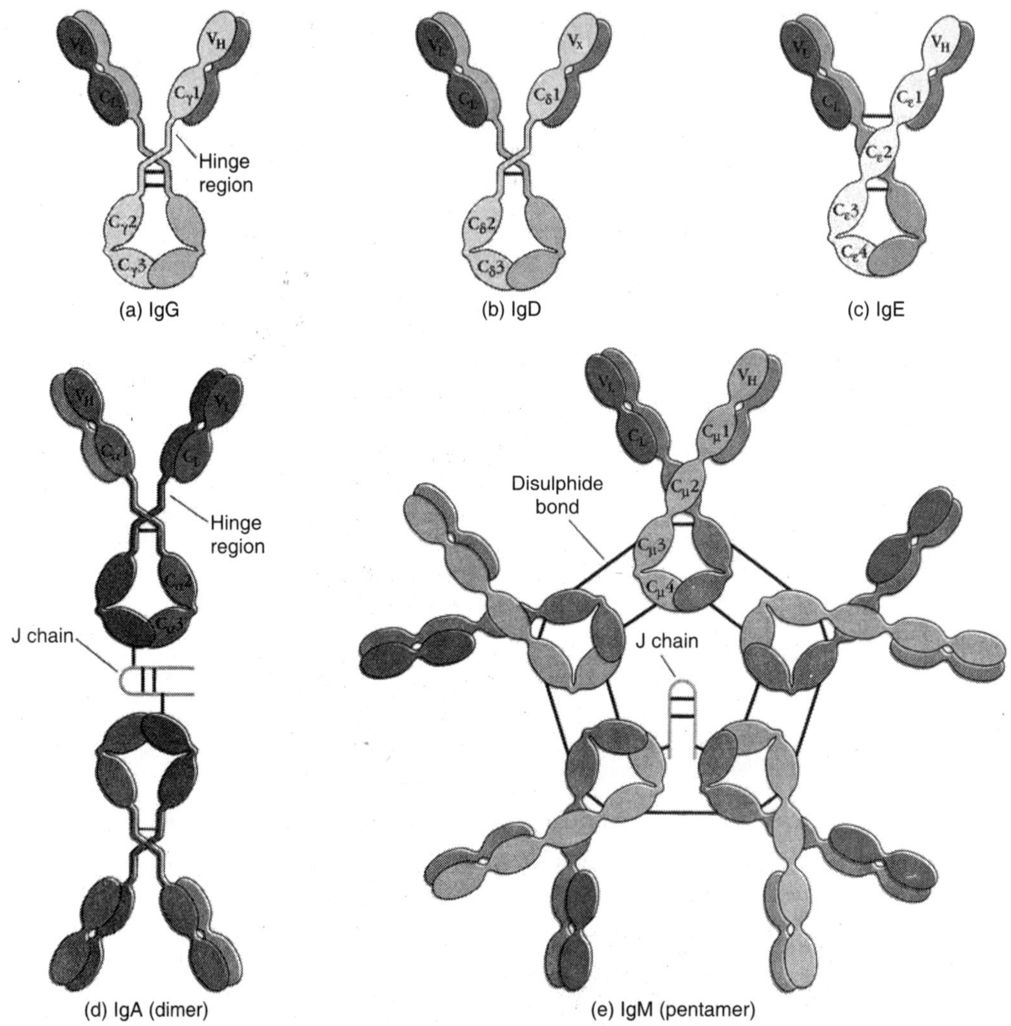

Figure 3.2 General structure of the five major classes of antibodies.

between the sub classes of IgG which affects the biological activity of this molecule. IgG1, IgG3 and IgG4 can cross the placenta and therefore, have an important role in protecting the developing foetus. IgG3 very effectively activates the complement followed by IgG1. IgG2 is a poor activator while IgG4 does not activate complement at all. IgG1 and IgG3 bind to Fc receptor on phagocytes very efficiently; IgG4 binds with an intermediate affinity, while IgG2 has an extreme low affinity. In an immune response IgG appears late and persists for longer duration.

It fixes complement, acts as opsonin and stimulates chemotaxis.

3.6.2 Immunoglobulin M

IgM is usually found as a pentamer of the basic four-chain unit. The μ chains have an extra constant domain as compared to IgG. The subunits of the pentamer are linked by disulphide bonds between the Cμ3 and Cμ4 domains. There is also an abundance of oligosaccharide units associated with the μ chain, and an additional peptide chain, the J (joining) chain. The J chain is cysteine-rich peptide of 137 amino acid residues and it assists the process of polymerization prior to secretion by the plasma cells. Each monomer contains two heavy (μ) chains and two light kappa (κ) or lambda (λ) chains. This molecule lacks the hinge region and hence, any flexibility.

IgM accounts for about 5–10 per cent of the immunoglobulin pool and is the major immunoglobulin class expressed on the surface of B-lymphocytes; IgM is largely confined to the intravascular pool. This is a pentamer with a molecular weight of 900,000 and a monomer weight of 180,000, and has an average concentration of 1.5 mg/ml in the serum. IgM has a half-life of five days. The presence of J chain allows the IgM to bind to receptor on the secretory cells, which transport it across the epithelial linings to the external secretions.

IgM is the first immunoglobulin class produced in a primary response to an antigen, and is also the first immunoglobulin to be synthesized by a neonate. The pentameric nature of IgM confers higher valency on the molecule, and it can bind multi-dimensional antigens like viral particles and RBC better. Complement activation requires two Fc regions in close proximity, and the pentameric structure of a single molecule of IgM fulfills this requirement.

Monomeric IgM is expressed as membrane bound antibody on B-cells and is used as an antigenic receptor by the B-cells. Pentamer of IgM is secreted by plasma cells in which five monomeric units are held together by disulphide bonds. The Fc regions of the monomers are directed towards the centre of the pentamer and the 10 antigen binding sites are directed to the periphery of the molecule. Each pentamer has got a J chain linked to the Fc region; it is bonded through disulphide bonds to two of the CH4 domains of two monomers. The J chain is added just before the formation of the pentamer and is required for the formation of the pentamer. These immunoglobulins cannot cross the placenta and are the first to appear in a primary response to an antigen and also the first one to disappear from the serum. It is a better agglutinating, complement fixing and bacteriolytic antibody. An IgM molecule can bind to ten small haptens at a time, but cannot bind to five large antigens because of steric hindrance.

This is a major component of rheumatoid factor (an autoantibody).

3.6.3 Immunoglobulin A

The amino acid residues of the α chain are arranged in four domains: V_H, Cα1, Cα2 and Cα3. IgA dimers are linked end to end through a J chain. Secretory IgA is made up of two units of IgA, one secretory component and a J chain. In contrast to the J chain, the **secretory component** is not synthesized by the plasma cell, but by epithelial cells. The secretory component protects the IgA from being degraded by enzymatic action.

This molecule represents 10–15 per cent of the human serum immunoglobulin pool and has a half-life of six days. In man, more than 80 per cent of the IgA occurs as monomer of the four-chain unit, but in most mammals the IgA occurs as a dimer. IgA is the predominant immunoglobulin in seromucous secretions such as saliva, tears, colostrum, breast milk and tracheobronchial and genitourinary secretions. The molecular weight of the monomeric molecule is 150,000. The dimeric IgA has a J chain polypeptide with a molecular weight of 15,000 and a polypeptide chain called the **secretory chain**. The J chain helps in polymerization of the serum IgA. The secretory component having a molecular weight of 70,000 is secreted by intestinal epithelial cells and hepatocytes. The secretory IgA (sIgA) exists mainly as a dimer and has a molecular weight of 385,000. Because IgA is polymeric, it can cross-link large antigens with multiple epitopes. Secretory IgA binds to bacterial and viral cell surface antigens, which in turn prevents attachment of these pathogens to the mucosal cells.

The daily secretion of IgA is far greater than any other immunoglobulin class. A normal human being secretes five to 15 gm of sIgA into mucus secretions every day. The polymeric IgA can bind to large antigens with multiple epitopes. The binding of sIgA to bacterial and viral surface antigens prevents them from binding to mucosal surfaces. Secretory IgA provides protection against bacteria like *Salmonella, N.gonorrhoes, Vibrio cholera* and viruses such as *polio virus, influenza virus* and *reoviruse*. IgA activates complement only through alternative pathway. It prevents adherence of microbes to mucus membranes.

3.6.4 Immunoglobulin D

Less than 1 per cent of total immunoglobulin in serum is IgD. There is a single disulphide bond between the δ chains and a large amount of carbohydrate distributed in multiple oligosaccharide units. In the heavy chain of IgD there are three N-linked polysaccharides and in the hinge region there is multiple O-linked polysaccharides.

Although IgD is a very small part of the total plasma immunoglobulins, it is present in large quantities on the membrane of many B-cells together with IgM and has the same antigen binding specificity as IgM. It has a possible role in the antigen-triggered lymphocyte differentiation. No biological effector function has been identified for IgD. It has a molecular weight of 150,000 and a half-life of three days. The serum concentration of IgD is about 0.03 mg/ml and the carbohydrate content is 13 per cent.

3.6.5 Immunoglobulin E

The ε chain has a higher molecular weight as it has a larger number of amino acid residues distributed over 5 domains—one variable and four constant domains. Although this class of immunoglobulins is scarce (0.004 per cent) in the serum, it is found on the membrane of basophils and mast cells in all individuals. It also sensitizes cells on the mucosal surfaces such as the conjunctival, nasal and bronchial mucosa. IgE is commonly associated with allergic diseases, such as asthma and hay fever. IgE molecules show a high affinity for binding to Fc receptors on mast cells through a site located on the Fc region of the IgE molecule. Cross linkage of receptor-bound IgE molecules by the allergen induces degranulation of basophils and mast cells, which results in the release of active mediators of allergy. The average concentration of IgE in the serum is 0.3 μg/ml. It has a half-life of 2.5 days and molecular weight of 190,000. The carbon content is 12 per cent and it is linked to the constant region of the heavy chain. The heavy chain of IgE bears six N-linked polysaccharides.

3.7 ANTIGENIC DETERMINANTS ON IMMUNOGLOBULINS

Differences in the amino acid sequences between Ig classes and subclasses determine the antigenic specificity of immunoglobulins (Figure 3.3). There are three major categories of antigenic determinants in immunoglobulins.

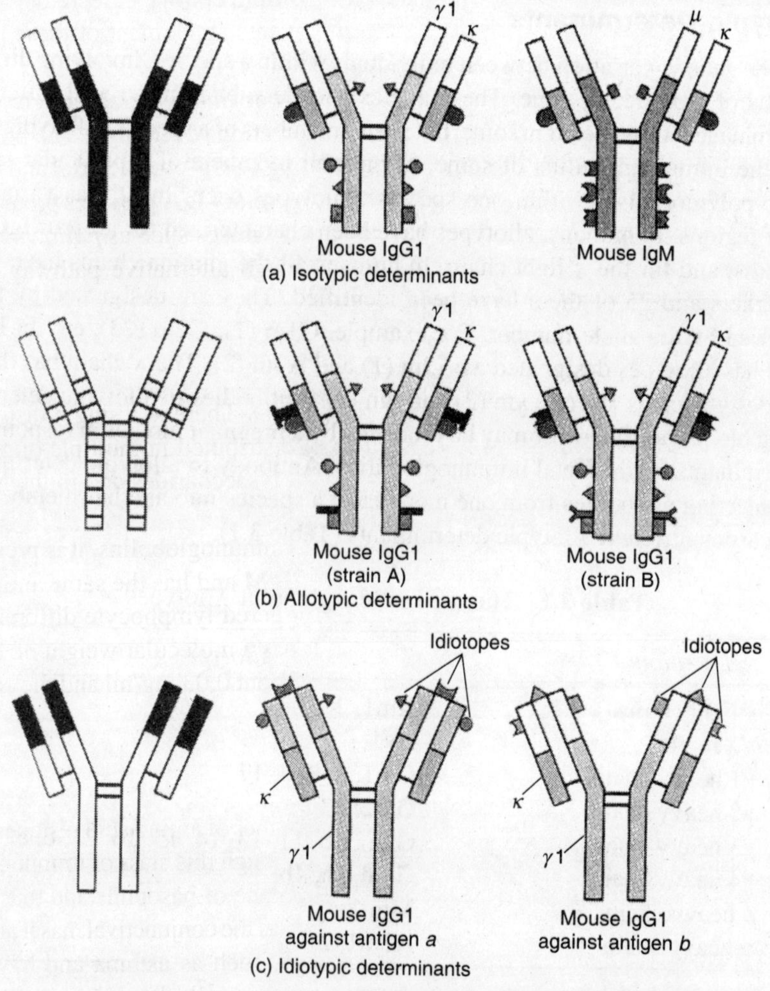

Figure 3.3 Antigenic determinants of immunoglobulins.

3.7.1 Isotypic Determinants

Variations observed among the heavy chains of different classes and subclasses of immunoglobulins are known as **isotypic variations**. The genes for the isotypic variations are present in all the healthy members of a species. For example, the genes for $\gamma 1$, $\gamma 2$, $\gamma 3$, $\gamma 4$, μ, $\alpha 1$, $\alpha 2$, δ, ε, κ and λ chains are all present in the human genome and are therefore isotypes. Each isotype has a distinct locus in the individual's genome. Isotypic determinants are constant region determinants that define each heavy-chain class and subclass and each light-chain type and subtype within a species. They are associated with the effector functions. Within a species each normal individual expresses all isotypes in the serum. Different species inherit different constant region genes and therefore, express different isotypes. Therefore, when an antibody from one species is injected into another species, the isotypic determinants will be recognized as foreign, and induces an antibody response to the isotypic determinants which are the constant region epitopes.

3.7.2 Allotypic Determinants

This refers to the genetic variation between individuals within a species, involving different alleles at a given locus of a particular gene. These alleles encode subtle amino acid differences called allotypic determinants, which occur in some, but not all members of a species. Allotypic determinants are found on the immunoglobulins of some, but not all members of a particular species. They reflect genetic polymorphism within one species. Allotypes occur mostly as variants of heavy chain constant regions. In humans, allotypes have been characterized for all four IgG subclasses, one IgA subclass, and for the κ light chain. In humans all the gamma chains have got allotypes called Gm markers and 25 of them have been identified. They are designated by the class and subclass followed by the allele number. For example, G1m (1), G2m (23), etc. In IgA, only the IgA2 subclass has allotypes designated as A2m (1) and A2m (2). The κ chain has three allotypic determinants designated as κm (1), κm (2) and κm (3). Antibodies to allotypic determinants may arise following blood transfusion or may be produced by a pregnant mother in response to paternal allotypic determinants in the foetal immunoglobulins. Antibody to allotypic determinants can be produced by injecting antibodies from one member of a species into another member of the same species, who carries different allotypic determinants (Table 3.1).

Table 3.1 Human Immunoglobulin Allotypes

Location	*Allotype*
κ light chains	Km1, 2, 3
V_H region	Hv1
γ1 heavy chain	Gm1, 2, 3, 4, 17
γ2 heavy chain	Gm2, 23
γ3 heavy chain	G3m5, 6, 11, 13, 14, 15, 16, 21, b, c
γ4 heavy chain	G4m, 4a, 4b
μ heavy chain	Mm 1
α heavy chain	A2m1, 2

Allotypes are used to identify people and show familial relationships. Possession of certain allotypic determinants is related to certain disease risks. Allotypes are not associated with ligand binding and effector functions. They are however important in probing the genetic basis for antibody structure and biosynthesis.

3.7.3 Idiotypic Determinants

Variation in the variable domain, particularly in the hypervariable regions, produces idiotypes. These are defined by their corresponding anti-idiotypes. The unique amino acid sequence of the V_H and V_L domains of a given antibody can function not only as antigen-binding site, but also as a set of antigenic determinants. These determine the binding specificity of the antigen-binding site. Idiotypes are usually specific for individual B-cell clones. The idiotypic determinants are generated by the conformation of heavy- and light-chain variable regions. They are a result of unique structures generated by hypervariable subregions on L and H chains. These are important for immune regulation. Each individual antigen determinant of the variable region is called as **idiotype**. The

α idiotypic determinant is the actual antigen binding site, while the β and γ idiotypes comprise the variable region sequences outside the antigen binding site. Each antibody presents multiple idiotypes and the sum of the individual idiotypes is called the idiotype of the antibody.

3.8 PRODUCTION OF IMMUNOGLOBULINS

Humoral immune response is initiated by antigen recognition and activation of mature or memory B-cells by CD4$^+$ T-cells. This results in the formation of plasma cells that synthesize and secrete antigen-specific immunoglobulin molecules.

Antigen-specific immunoglobulin molecules are synthesized in two structurally similar forms:

- **B-cell receptors (BCRs) or surface receptors (sIgs):** These immunoglobulins are secreted by B-lymphocytes and bound to their surface where they function as antigen receptors (binding antigens at the variable region of the sIg).
- **Antibody molecules:** These are synthesized and secreted into body fluids by plasma cells of the B-cell lineage.

3.8.1 Factors Affecting Antibody Production

The nature and the extent of immunoglobulin production during an antibody response depends mainly on:

(i) Function of the immune system, e.g. presence of various cytokines and accessory cells, function of helper T-cells and their cytokines.
(ii) Nature of the antigen, its dose and site of entry.
(iii) First or second exposure of the host to the particular antigen.

Primary immune response

Primary immune response is characterized by a production of antibodies on first exposure to a particular antigen. As a B-lymphocyte encounters an antigen in the lymphoid tissue, it becomes activated and may either differentiate into effector cell (antibody secreting cell) or become a B-memory cell. **The effector B-cells** in a primary response mainly synthesize IgM molecules, followed by IgG molecules that show specificity for the stimulating antigen. Their function as effector molecules depends on their ability to participate in an immune response either by inactivating or eliminating the stimulating antigen.

Secondary immune response

A second encounter with the same antigen showing the same specificity as the B-cell receptor on the surface of a B-memory cell triggers a secondary immune response also known as anamnestic response. The antibodies produced in the secondary response are mainly of the IgG class. The secondary response is much greater than the primary response and shows higher affinity towards the stimulating antigen (Figure 3.4, and Table 3.2).

Antibodies generally do not kill or remove pathogens solely by binding to them. To be effective against pathogens, antibodies must recognize the antigen and also invoke immune effector functions that will remove the antigen and kill the pathogen. The variable regions of the antibody are responsible for binding to antigen; the heavy chain constant region is responsible for a variety of interactions with other proteins, cells and tissues that result in the effector functions of the humoral response.

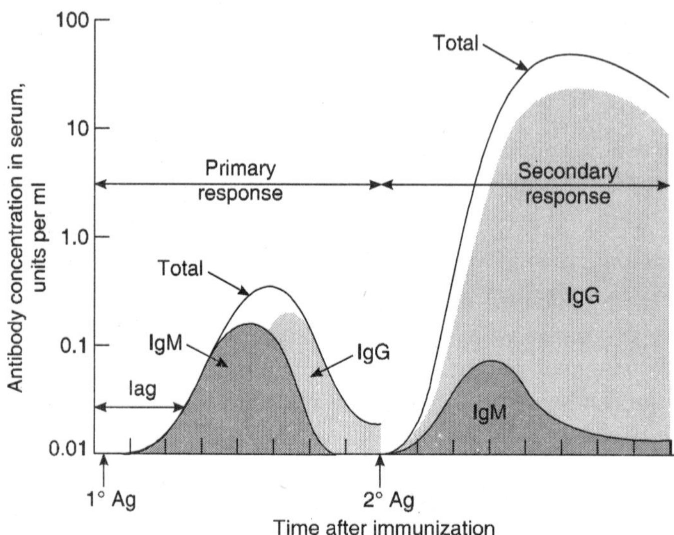

Figure 3.4 Concentration and isotype of serum antibody following primary and secondary immunization with antigen.

Table 3.2 Comparison of Primary and Secondary Immune Responses

Property	*Primary Response*	*Secondary Response*
Responding B-cell	Naive (virgin) B-cell	Memory B-cell
Lag period following antigen administration	Generally 4–7 days	Generally 1–3 days
Time of peak response	7–10 days	3–5 days
Magnitude of peak antibody response	Varies depending on antigen	Generally 100–1000 times higher primary response
Isotype produced	IgM predominates early in the response	IgG predominates
Antigens	Thymus-dependent and thymus-independent	Thymus-dependent
Antibody affinity	Lower	Higher

There are three major effector functions that enable antibodies to remove antigens and kill pathogens: **Opsonization** promotes antigen phagocytosis by macrophages and neutrophils. **Complement activation** by IgM and most human IgG classes activates a pathway that can perforate cell membranes. **Antibody-Dependent Cell-mediated Cytotoxicity (ADCC)** causes NK-mediated death of target cells when antibody bound to the target cells associates with the Fc receptors of the NK-cells.

3.9 ANTIBODY DIVERSITY

The immune system is essential for survival. Without it, death from infection would be inevitable. The molecules of the immune system maintain a constant surveillance against invading microorganisms. Antibodies are remarkably diverse and provide enough different combining sites to recognize the millions of antigenic shapes in the environment. Each class of antibody also has a characteristic effector region so that they can bind to Fc receptors. An individual produces more different forms of antibody than all the other proteins of the body together (Figure 3.5). In fact more types of antibodies are produced than there are genes in our genome.

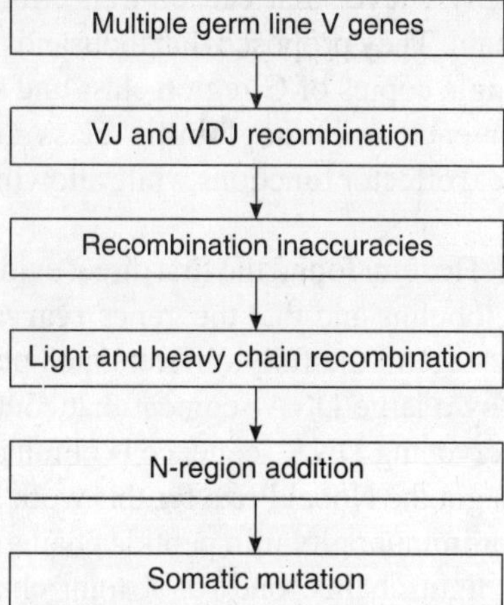

Figure 3.5 The methods of generating antibody diversity.

Antibodies recognize a limitless variety of foreign substances and can distinguish and bind to every possible antigen that the host is likely to meet during its lifetime. They also remember each infection so that the second exposure to the same organism is dealt with more efficiency. This implies that a limitless variety of antibody molecules need to be synthesized. An antibody is an assembly of protein chains and a gene specifies its structure. To solve the paradox of a limited number of genes and an apparently limitless capacity to generate different antibodies two competing theories were put forward in explanation: the germ line and somatic recombination theories.

Immunoglobulins are composed of heavy and light chains, the light chains being either kappa or lambda. Since any light chain can combine with any heavy chain, the number of possible antigen binding sites is the product of the number of heavy and light chains.

3.10 IMMUNOGLOBULIN GENE STRUCTURE

Any viable model of the immunoglobulin genes have to account for the following properties of antibodies:

1. The vast diversity of antibody specificities.
2. The presence of the variable region of heavy and light chains at the amino terminal end and a constant region at the carboxyl terminal end.

3. The existence of isotypes with the same antigenic specificity, which is generated from the association of a specific variable region with different heavy chain constant regions.

The germ line theories maintain that the genome of the germ cells have a large repertoire of immunoglobulin genes. On the other hand, **the somatic variation theories** say that the genome contains a relatively small number of immunoglobulin genes from which a large number of antibody specificities are generated in the somatic cells by mutation or recombination.

In 1965, W. Dreyer and J. Bennett proposed a **two-gene one-polypeptide model**. They suggested that two separate genes encode a single immunoglobulin heavy or light chain, one gene for the variable region and the other for the constant region. They suggested that these two genes somehow come together at the DNA level that can be transcribed and translated into a single immunoglobulin heavy or light chain. They proposed that thousands of V-region genes were carried in the germ line, whereas only single copies of C-region class and subclass genes need to exist. A single constant region gene for each immunoglobulin class and subclass accounts for the conservation of necessary biological effector functions while allowing for diversification of variable region genes.

In 1976, S. Tonegawa and N. Hozumi found the first direct evidence that separate genes encode the V and C regions of immunoglobulins and that the **genes rearranged** during differentiation of lymphocytes from the embryonic stage to the fully differentiated plasma cell stage. In the embryo, the V and C genes are separated by a large DNA segment that contains a restriction endonuclease site; during differentiation the intervening DNA sequence is eliminated and the V and C genes are brought close together. Tonegawa got the Nobel Prize for this work in 1987.

Three gene families code for immunoglobulin peptide chains and they are found on different chromosomes. One codes for κ light chains, one for λ light chains and one for heavy chains (Figure 3.6). Each of these three gene families contains a large number of gene segments coding for variable regions (V genes), one or more gene segments coding for constant regions (C genes) and a variable number of very small gene segments known as J (joining) and D (diversity) segments. Antibody diversity arises as a result of recombination between these gene segments.

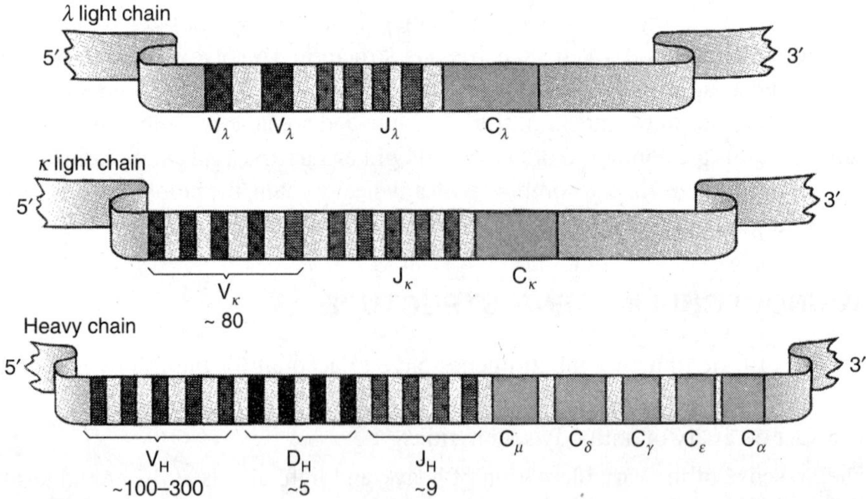

Figure 3.6 The structure of genes for lambda light chains, kappa light chains and the heavy chains. Each is located at a different chromosome.

3.10.1 Immunoglobulin Gene Recombination

The kappa and lambda light chain families contain V, J and C genes segments; the rearranged VJ segments encode the variable region of the light chains. The heavy chain family contains V, D, J and C genes segments; the rearranged VDJ genes segments encode the variable region of the heavy chain. In each gene family, C gene segments encode the constant regions. Every V gene segment has got a small exon at its 5′ end that encodes for short signal or leader peptide.

In humans, the Lambda locus has 30 functional Vλ genes segments, four functional Jλ segments and seven Cλ segments, only four of which are functional. In addition to the functional genes segments, the human Lambda complex contains many Vλ and Jλ pseudogenes.

The kappa chain multigene family in humans contains approximately forty Vκ gene segments, five Jκ segments and a single Cκ segment. Since there is only one Cκ gene segment, there are no subtypes of κ light chains.

Light chain genes recombine V and J gene segments to make a gene for the V domain. The heavy chain variable region is also encoded by V and J segment genes. A third gene segment, the D, diversity segment gene, provides additional diversity. The D segment is highly variable for in the number of codons and in the sequence of the base pairs. Thus Heavy chain genes recombine V, D and J gene segments to make a gene for the V_H domain. V regions are rearranged and expressed in a programmed manner during early foetal life. The place at which V and J segment genes join may also vary slightly.

Two distinct genes code for each immunoglobulin peptide chain, one for the variable region and one for the constant region. Therefore, a single constant region gene could be combined with any one of many variable region genes. Since multiple genes code for immunoglobulins it is necessary to store only the information about all the variable regions and to match these with a smaller number of constant region genes in order to produce a complete range of immunoglobulin molecules.

3.10.2 Generation of V Region Diversity

To generate a great variety of antibodies with different specificity for antigen, it is necessary to generate great diversity in the amino acid sequence in the variable regions of both light and heavy chains. The most obvious mechanism of generating V region diversity is to randomly select one V gene segment from the available pool and join it to a randomly selected J gene segment. Since there are multiple V and J gene segments available, the number of possible combinations can be very large.

V(D)J joining is catalyzed by the **recombination activating genes** RAG-1 and RAG-2, and the participation of other enzymes and proteins. The joining of segments is directed by Recombination Signal Sequences (RSSs), which are conserved DNA sequences that flank each V, D, and J gene segment. Each recombination signal sequence contains a conserved heptamer sequence, a conserved nonamer sequence and either a 12 base pair (one turn) or 23 base pair (two turn) spacer. During rearrangement, gene segments flanked by one-turn spacer join only to segments flanked by a two-turn spacer, which ensures proper V_L–J_L and V_H–D_H–J_H joining.

In gene rearrangement the heavy chain rearrangement occurs first which is followed by light chain rearrangements. The functional rearrangement of the immunoglobulin DNA of only one parental chromosome results in allelic exclusion, which is necessary to ensure that a mature B-cell expresses immunoglobulin with a single antigenic specificity. Antibody diversity can generate more than 10^{10} possible antibody combining sites. This diversity is the result of random joining of multiple V, J and D germ line genes segments: P-addition; N-addition; and the somatic mutation.

Joining position

The actual base at which V and J gene segments join can vary and thus changes in the base sequence occur at the splice site around position 96. This variability in V–J joining is important in generating different amino acid sequences in this region of the light chain.

Base insertion

Random bases may be inserted or deleted at the V–D and D–J splice sites. This process is called N-region addition. Some bases may also be removed from the exposed ends.

Somatic mutation

Spontaneous genetic change is known to occur in developing cells. Mutations are seen in the variable domain genes and in the immediately adjacent regions, but not in the constant domains. Rates of mutation suggest that there should be one change in the V region for every 3 to 30 cell divisions. The somatic recombination gives rise to cells that vary greatly in their antigen specificity.

Since both light and heavy chains CDRs contribute to the antibody-binding site, the number of potential combinations and therefore, binding specificity goes up.

3.10.3 Constant Region Diversity

The constant region genes for immunoglobulin heavy chains are selected without regard to V region diversity. As a result, any individual V region may be associated with any one of the five immunoglobulin class constant regions. Class switching which occurs after antigenic stimulation of mature plasma cells also contributes towards constant region diversity. This results in expression of different classes of antibodies with the same antigenic specificity.

Thus, the immune system is able to recognize and respond to many antigens by generating great diversity in the antibodies produced by the B-cells and in the antigen receptors expressed by T-cells. Diversity is achieved by the recombination of a limited number of V, D and J gene segments to produce a vast number of the variable domains. Immunoglobulin heavy and light chains undergo structural modifications after antigen stimulation, called somatic mutation. In addition to simple combinations of V, D, and J gene regions, diversity in the TCR and immunoglobulins depend upon N region diversification, joining site variation and multiple D regions.

After antigenic stimulation of mature B-cells, class switching occur which results in the expression of different classes of antibody (IgG, IgA and IgE) with the same antigenic specificity. Immunoglobulin class switching involves recombination of the VDJ genes with various C region genes and differential RNA splicing.

MULTIPLE-CHOICE QUESTIONS

1. The number of constant domains in the heavy chain of IgE is:
   (a) one (b) two
   (c) three (d) four

2. The antigen combining site of an antibody molecule determines its:
   (a) isotype (b) allotype
   (c) idiotype (d) anti-idiotype
3. The immunoglobulin that activates complement upon binding with antigen is:
   (a) IgM (b) IgD
   (c) IgA (d) IgE
4. Molecules in the immunoglobulin superfamily share:
   (a) antigen binding sites (b) domains
   (c) variable regions (d) tyrosine residues
5. The complementarity determining regions of immunoglobulins:
   (a) activate complement (b) bind to antigen
   (c) bind to cells (d) mediate opsonization
6. The strength of binding between an antigen and its antibody is called:
   (a) affinity (b) avidity
   (c) valency (d) hydrophobicity

REVIEW QUESTIONS

3.1 What are the differences between isotypes, allotypes and idiotypes?

3.2 Give the general structure of immunoglobulin.

3.3 Give a detailed account of gene rearrangements in variable regions.

3.4 What are primary and secondary immune responses?

3.5 Which genes code for immunoglobulin light chain?

3.6 Which are the two immunoglobulin classes that can be expressed simultaneously on a B-cell surface?

3.7 How does immunoglobulin class switch occur?

3.8 What is N region addition?

ANTIGEN–ANTIBODY INTERACTION

4.1 ANTIGEN–ANTIBODY INTERACTION: AN OVERVIEW

Antibody molecules combine reversibly with antigens to form immune complexes. When an antigen and its specific antibody combine, they interact through the chemical groups on the surface of the epitope and the paratopes of the immunoglobulin molecule. In other words, in the reaction between antigen and antibody the antigenic determinants interact with the V_H/V_L domain of the antibody molecule particularly the hypervariable region. This interaction is through non-covalent bonds. The non-covalent interactions that occur between the antibody and antigen are hydrogen bonds, van der Waals interactions, ionic bonds and hydrophobic interactions.

Each type of non-covalent bond is relatively weak in itself, but collectively the bonds may have a significant binding strength. Strong binding occurs when the two components are closely placed and when the shape of the epitope and the shape of the CDRs conform to each other. The strongest binding between an epitope and immunoglobulin CDRs occurs only if their shapes match perfectly so that multiple non-covalent bonds can form. However, antigens can bind to antibodies when they fit less than perfectly, but the number of bonds established and the strength of binding will be less.

4.2 ANTIGEN–ANTIBODY COMPLEX

Close proximity of the interacting molecules is an important factor in the formation of antigen-antibody complex using such binding forces as ionic interactions, hydrogen bonds, hydrophobic bonds, and van der Waals forces. Environmental factors like hydrogen ion concentration (pH), temperature, and salt concentration (ionic strength) also affect antigen-antibody interactions.

Maximum binding of antigen by an antibody occurs only in the presence of such conditions as:

1. Complementarity (specificity) of the two interacting molecules.
2. Close proximity of antigen to its specific antibody.
3. High affinity and avidity of the antibody for the particular antigen.
4. Presence of approximately equal number of multivalent sites.
5. Appropriate concentration of antigens and antibodies (zone of equivalence).
6. Appropriate environmental factors.
7. The greater the number of binding sites, the larger is the resulting antigen-antibody complex.
8. The size of an antigen-antibody complex depends largely on the valency of the antigen. When an antibody interacts with a bivalent antigen, only linear chains or circular complexes are produced.

When an antigen-antibody reaction occurs, the reacting molecules are of the same specificity and all other factors that support binding of an antigen with its specific antibody are appropriate for the reaction. The reaction may be represented as:

$$Ag + Ab = Ag/Ab$$

where Ag represents an antigen, Ab is the antibody and Ag/Ab is the formed complex.

The stability of an antigen-antibody complex depends on the forces that exist between the two molecules.

Three groups of techniques are used to detect and measure antigen-antibody combination. The most sensitive technique is to directly measure the immune complexes formed in an in-vitro system. These tests are known as **primary binding tests**.

4.2.1 Primary Antigen–Antibody Interaction

Primary antigen-antibody interaction represents the first stage of binding between an epitope and the antigen-specific binding site on an antibody molecule. The initial binding depends on forces of attraction (affinity and avidity) between these molecules. An optimal ratio of antigen and antibody molecules produces maximal conditions for binding of an antigen with its complementary antibody and this is referred to as the *zone of equivalence*. Primary binding tests are performed by allowing antigen and antibody to combine and then measuring the amount of immune complex formed. Radioisotopes, fluorescent dyes, or enzymes are used as labels to identify one of the reactants.

4.2.2 Secondary Antigen–Antibody Interaction

Secondary antigen-antibody interaction occurs between multivalent antigens and antigen specific antibodies. These interactions depend on cross-linking of various antigen molecules by specific antibodies to produce lattices or aggregates that can be seen as precipitation or agglutination.

4.2.3 Affinity and Avidity

The strength of the total non-covalent interactions between a single antigen-binding site on an antibody and a single epitope is the **affinity** of the antibody for that epitope. Affinity refers to an intrinsic force of attraction between antigen binding site on an antibody and one epitope on a corresponding univalent antigen. Low affinity antibodies bind the antigen weakly and tend to dissociate readily, whereas high affinity antibodies bind antigen more tightly and remain bound longer. The affinity at one binding site does not always reflect the true strength of the interaction between antibody and antigen. When complex antigens containing multiple, repeating antigenic determinants are mixed with antibodies containing multiple binding sites, the interaction of an antibody molecule with an antigen molecule at one site increases the probability of reaction between those two molecules at a second site. The strength of such multiple interactions between multivalent antibody and antigen is called **avidity**. The avidity of an antibody molecule is the better measure of its binding capacity within biological systems than the affinity of its individual binding sites. High avidity can compensate for low affinity. The avidity is overall binding capacity between antibodies and multivalent antigens and is a measure of the stability of an antigen-antibody complex.

After the administration of antigen, the average affinity of the IgG antibodies produced gradually increases as an immune response progresses. The affinity of IgG also tends to rise as the dose of immunizing antigen is lowered. The affinity of IgM antibodies remains unchanged. It is thought that a high initial dose of antigen can provoke many clones of antibody producing cells with high, medium, or low affinity for antigen. As an immune response progresses, the amount of available antigen is gradually reduced, which tends to stimulate only those cells which have a high affinity for the antigen. The antibodies produced late in an immune response will therefore, have a relatively high average affinity. In general, high affinity antibodies are less specific for antigen than low affinity antibodies. High affinity antibodies combine strongly not only with their inducing, but also with related epitopes, for which their affinity may be much lower. In contrast, low affinity antibodies react with their inducing epitope, but are unlikely to react detectably with any other epitope; thus giving the appearance of high specificity. High affinity antibodies are superior to low affinity antibodies for virus and toxin neutralization. They are also more effective at removal of foreign antigen from the blood stream and at complement activation and red blood cell destruction through haemolysis.

4.2.4 Cross-reactions

Although Ag-Ab reactions are highly specific, sometimes antibody elicited by one antigen can cross-react with an unrelated antigen. Cross-reactivity refers to an immunologic reaction in which specific antibodies react with the following:

(i) Two dissimilar antigens that share common epitopes.
(ii) An epitope with a structural resemblance to the epitope that stimulated their production.
(iii) Heterophil antigen (from other species such as animals, microorganisms, plants) that is structurally identical or closely related. Specific antibodies that cross-react with heterophil antigens are known as **heterophil antibodies**.

Once in a while certain antigens show **cross-reactivity** with unrelated antibodies. This is possible only when two different antigens share similar epitopes, or antibodies specific for one epitope also bind to an unrelated epitope possessing similar chemical properties. Different types of viruses and bacteria are known to possess antigenic determinants that resemble normal host cell antigens. In some cases these antigens of viral or bacterial origin induce antibody production in the host, which cross-react with host cell antigens. Antibodies to duck ovalbumin cross-react with chicken ovalbumin since both these two proteins are very similar in structure.

4.3 BIOLOGICAL CONSEQUENCES OF ANTIGEN–ANTIBODY INTERACTION

The combination of antigen and antibody may block toxic sites on toxins or prevent viruses from binding to cells. The role of antibodies is to identify the target antigen. It is only through reactions mediated by constant regions of antigen-bound immunoglobulins that major protective mechanisms are triggered. Antibody molecules are therefore, transducers that convert the initial antigen-antibody interaction into a protective response. After combining with antigen, antibody can activate the complement cascade, bind effectively to phagocytic cells or provoke the degranulation of mast cells and basophils. The most significant effect is the triggering of the antibody response itself, when antigen binds to cell membrane immunoglobulins on B-lymphocytes. These reactions are initiated by a conformational change that occurs in the immunoglobulin Fc region as a result of antigen combining with the Fab region, and the result is thought to *form new, biologically active sites*. Alternatively, the *active sites preexist hidden* in the uncomplexed antibody molecule. When antigen binds to the antibody the resulting conformational changes result in exposure of these hidden sites.

Another type of cellular response to antigen-antibody-cell membrane interaction is *degranulation*. This occurs in mast cells, platelets, and neutrophils and makes the cell release the contents of its granules into the extracellular fluid.

Cytotoxic activity may also be generated in response to immune complex formation. Cells that can participate in this process (known as antibody dependent cellular cytotoxicity or ADCC) and destroy foreign cells on contact include macrophages, neutrophils and some lymphocytes subpopulations. These cells exert their cytotoxicity regardless of whether they first bind to antibody-coated target cells or the antibody and then bind to antigens on target-cell surfaces.

4.4 PRECIPITIN REACTIONS

Interaction between an antibody and soluble antigen in aqueous solution forms a lattice that eventually develops into a visible precipitate. Antibodies that bring about precipitate formation on reaction with antigens are called **precipitins**. Formation of an antigen-antibody lattice depends on the valency of both antigen and antibody. To form a visible precipitate the antibody should bivalent and the antigen should be bivalent or polyvalent. The antibody must be bivalent; a precipitate will not be formed with monovalent Fab fragments. The antigen must be either bivalent or polyvalent; it must have at least two copies of the same epitope or have different epitopes that react with different antibodies present in polyclonal antisera.

If increasing amount of soluble antigen are mixed with a constant amount of antibody, the amount of precipitate formed depends on the relative proportions of the reactants. No precipitate is formed at fairly low antigen concentrations. As the amount of antigen added increases, a precipitate forms and increases in amount until it reaches a maximum. With the addition of even more antigen, the amount of precipitate begins to diminish, until none is observed in tubes containing a large excess of antigen over antibody. Although formation of the soluble antigen-antibody complex occurs within minutes, formation of the visible precipitate occurs more slowly and often takes a day or two to reach completion. This formation of insoluble antigen-antibody complexes when a soluble antigen is mixed with its corresponding antibody is called **immunoprecipitation**. Formation of these visible precipitins occurs at the *zone of equivalence*, when the antigen concentration (number of available multivalent antigen sites or epitopes) and the antibody (number of antigen binding sites on the specific antibodies) concentration are approximately equal. More specifically, the number of epitopes is equal to the number of antigen binding sites in the equivalence zone.

4.4.1 Precipitation Reactions in Fluids

A quantitative precipitation reaction is performed by placing a constant amount of antibody in a series of tubes and adding increasing amounts of antigen to them. The precipitate formed in each tube is measured. Plotting the amount of precipitate against increasing antigen concentration yields a precipitin curve. Excess of either antibody or antigen interferes with maximal precipitation, which occurs in the so-called equivalence zone, when the ratio of antibody to antigen is optimal.

As increasing concentrations of antigen are added to a constant amount of antibody, the amount of immune complex precipitate rises and then falls. The precipitin curve generated in this way has three zones (Figure 4.1).

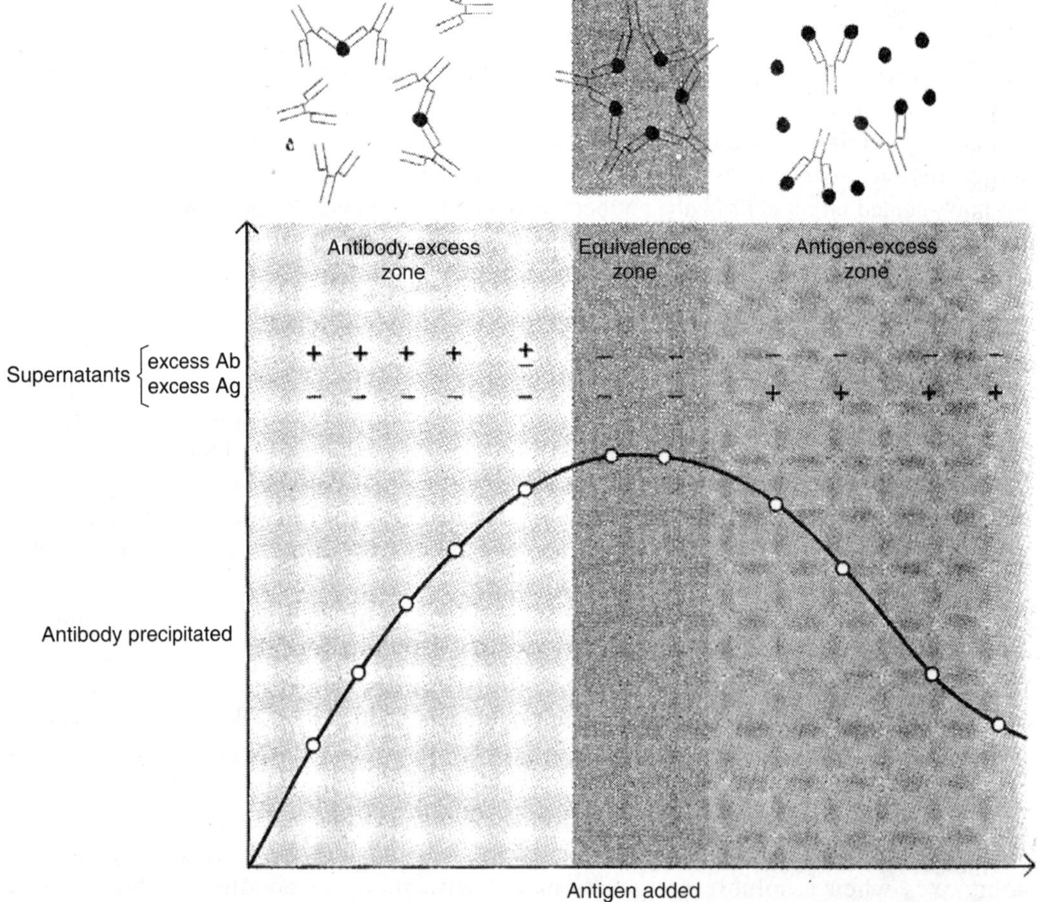

Figure 4.1 A precipitation curve for a system of one antigen and its antibodies.

Antibody excess zone

When the antibody is in excess, unreacted antibody will be found in the supernatant. Since the amount of antigen is insufficient to react with and precipitate the entire antibody present; thus free antibody can be detected in the supernatant. This zone contains soluble complexes consisting of multiple molecules of antibody bound to a single molecule of antigen. Each antigen molecule is effectively covered with antibody, which prevents cross-linkage and thus, precipitation.

Equivalence zone

When the antigen-antibody ratio in the immune complex is optimal, maximum precipitate will be formed which is called the equivalence zone. The added antigen is sufficient to combine with and precipitate the entire antibody present. Therefore, neither free antigen nor antibody can be detected in the supernatant. Maximal precipitation occurs at equivalence zone, where the ratio of antigen to antibody is optimal. As a larger multimolecular lattice is formed in this zone, the complex increases in size and precipitates out of solution.

Antigen excess zone

In the antigen excess region, unreacted antigen will be found in the supernatant with one or two molecules of antigen bound to a single molecule of antibody. Here the amount of antigen exceeds the amount that is required to bind the entire antibody. This leads to a reduction in the amount of antibody precipitated. Small soluble complexes are seen consisting of one or two molecules of antigen bound to a single molecule of antibody. Further cross-linkage is impossible in this case because of insufficient antibody, and since these complexes are small and soluble, no precipitation occurs. The extent to which this phenomenon occurs varies with different antibodies and with the species from which the antibody is derived.

In both antigen-excess and antibody-excess zones, precipitation is inhibited because extensive lattices cannot be formed.

Titration of antibodies

The amount of specific antibodies is often determined by titration. Titration is a procedure in which the serum under test is made up into a series of increasing dilutions. Each dilution is tested for activity in the test system. The reciprocal of the highest dilution giving of positive reaction is known as the **titer**, and provides a measure of the amount of antibody in that serum.

4.4.2 Precipitation Reactions in Gels

Immune precipitates can form in agar gels too. When antigen and antibody diffuse towards one another in agar (double diffusion or Ouchterlony method) or when antibody is incorporated into the agar and antigen diffuses into the antibody-containing matrix (radial immunodiffusion or Macini method), a visible line of precipitation will form. This visible precipitation occurs in the region of equivalence. No visible precipitate forms in the regions of antibody or antigen excess. These immunodiffusion reactions are used to determine relative concentrations of antibodies or antigens, to compare antigens, or to determine the relative purity of an antigen preparation. Two immunodiffusion techniques regularly used are **double immunodiffusion** and **radial immunodiffusion** (Figure 4.2).

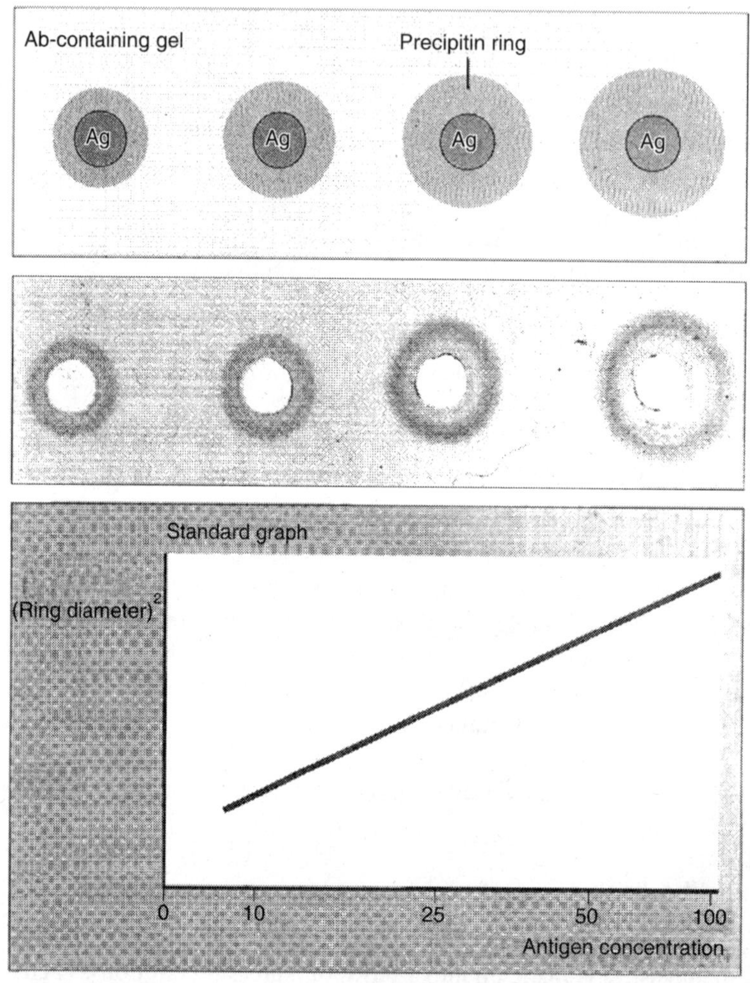

Figure 4.2 Quantitative analysis of antigens by single radial immunodiffusion.

Double immunodiffusion

In **double immunodiffusion**, molten agarose is prepared in a suitable buffer and poured on a slide and allowed to solidify. Wells are punched on the gel and unknown samples of antigens are tested for precipitin reaction against an antibody in the middle well. The principle of this technique is diffusion of antigen and antibody radially from wells towards each other through a concentration gradient. When equivalence is reached a white precipitin arc appears at the junction of antigen-antibody complex.

Radial immunodiffusion

In **radial immunodiffusion**, the gel contains a suitable dilution of antibody; antigen is added to the wells and allowed to diffuse through the agarose. As the antigen diffuses into the agarose a ring of precipitate is formed in the region of equivalence. The diameter of the precipitin ring is proportional to the concentration of antigen. By preparing a standard curve by measuring the diameter of rings formed against known concentration of antigen against a known dilution of antibody, one can find out the concentration of unknown antigen.

4.5 IMMUNOELECTROPHORESIS

The technique of **immunoelectrophoresis** combines separation by electrophoresis with identification by double immunodiffusion (Figure 4.3). Components of an antigen mixture are first separated on the basis of charge by electrophoresis. Troughs are then cut in the agar gel parallel to the direction of the electric field, and antiserum added to the troughs. Antibody and antigen then diffuse towards each other and produce lines of precipitation where they meet in appropriate proportions. Immunoelectrophoresis is a *qualitative* technique that can detect relatively high antibody concentration. Immunoelectrophoresis is used for checking the purity of isolated serum components like antigens or antibodies.

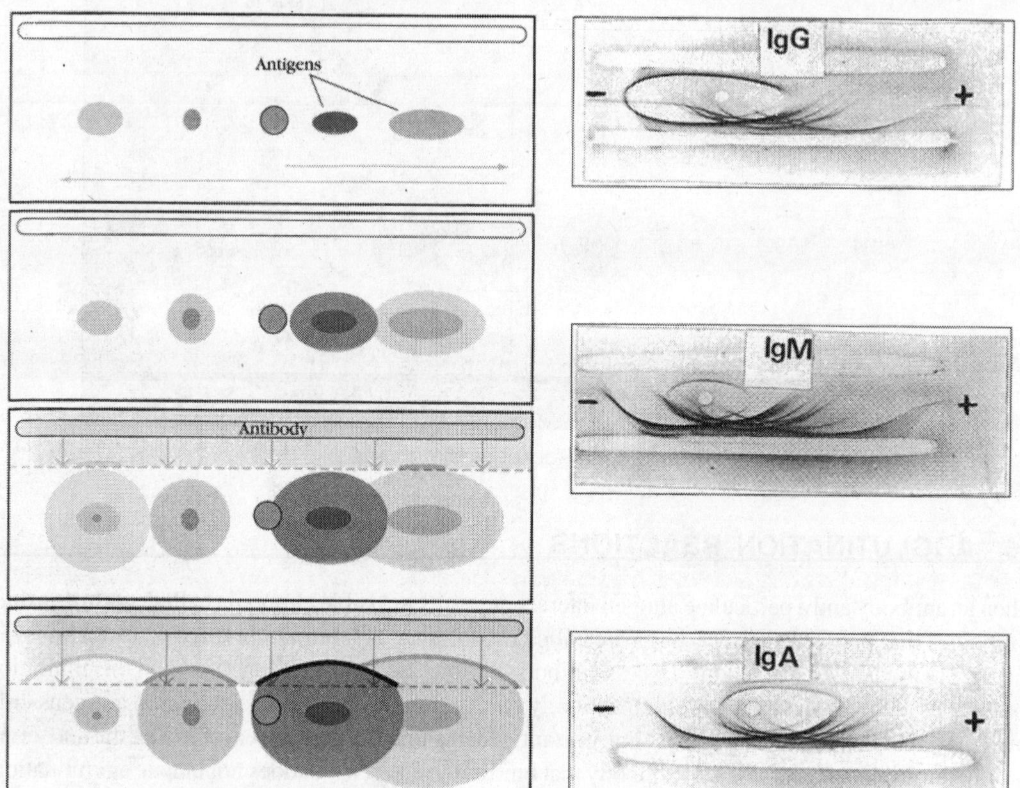

Figure 4.3 Immunoelectrophoresis of an antigen mixture.

A related *quantitative* technique, **rocket electrophoresis** can measure antigen levels. The gel contains the antibody of known concentration and negatively charged antigen is loaded in the wells punched in one corner of the gel plate. The antigen well side is connected to the negative pole and the opposite end to the positive pole. Electrophoresis is carried out and as antigen starts travelling in the gel towards the positive pole, precipitin band appears in the form of a rocket. The precipitate formed between antigen and antibody is in the shape of a rocket, the height of which is proportional to the concentration of antigen in the well (Figure 4.4). A limitation is that only negatively charged antigen is able to move in this matrix, and it is not possible to quantitate several antigens in a mixture at the same time.

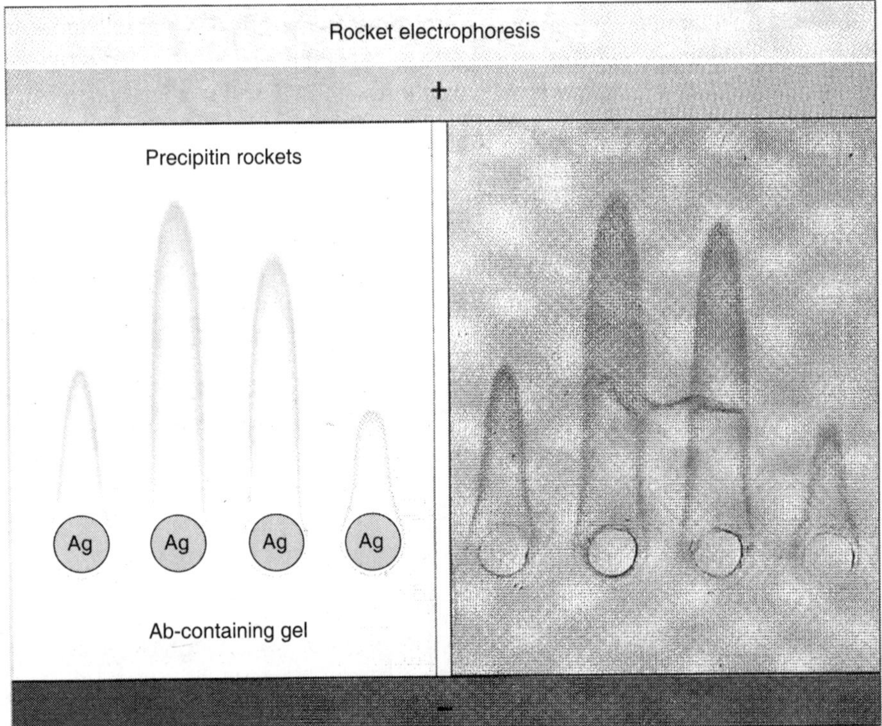

Figure 4.4 Rocket electrophoresis.

4.6 AGGLUTINATION REACTIONS

When an antibody and a particulate antigen interact, it results in visible clumping called **agglutination**. Antibodies that produce such reactions are called agglutinins. Agglutination reactions are similar in principle to precipitation reactions. Excess antibody inhibits agglutination reactions and this inhibition is known as the **prozone effect**. At high antibody concentration, antibodies can bind to antigens only univalently and therefore, cannot cross link one antigen to another. Another reason is that the antiserum may contain high concentrations of antibody that bind to the antigen, but does not induce agglutination. Such antibodies are called **incomplete antibodies** and are often of the IgG class. At high concentrations, these incomplete antibodies (IgG) may occupy all the antigenic sites and block access by IgM, which is a good agglutinin. Alternatively, the density of epitope distribution, or the location of some epitopes in deep pockets of a particulate antigen, may make it difficult for the antibodies specific for these epitopes to agglutinate certain particulate antigens.

Haemeagglutination

When agglutination reactions are done with RBC then they are termed as **haemeagglutination**. Typing of ABO blood group is a hemeagglutination reaction. Agglutination with an antiserum of A or B would indicate to which blood group the person belongs. Many bacteria can be identified by agglutination reactions.

The presence of non-agglutinating antibodies can be tested by direct or indirect antiglobulin test.

Coombs test

Sometimes antibody molecules directed against deeply embedded membrane antigenic determinants cannot cause agglutination even after binding to antigenic determinants. Such antibody molecules are short because their Fab arms cannot bridge antigenic determinants on two different cells. Therefore, agglutination fails to occur even after the antibody molecules binds to the antigenic determinants. Rh antibodies are of this category. These are called as incomplete antibodies because they fail to bring about agglutination of RBCs even after binding of antibody molecules to the red cells carrying Rh antigen. These antibody bound RBC can be agglutinated by the addition of anti-human antibodies. This is called Coombs test which detects the presence of incomplete antibodies.

Passive agglutination

Sometimes a precipitating system has to be converted into an agglutinating one, because agglutination is more sensitive technique than precipitation. This is done by chemically linking soluble antigen to inert particles, like red blood cells or bacteria, so that specific antibody will make the sensitized particles agglutinate. Red blood cells are the best particles for this purpose, tests that employ coated red blood cells are called **passive haemagglutination**.

Particulate surfaces used here may be tanned and killed RBCs, particles of latex or polystyrene, bentonite, etc. RBCs are tanned by treating them with tannic acid or chromium chloride which enhances the adsorption of the antigen to the surface of the RBC.

4.6.1 Agglutination Inhibition Reaction

These reactions are used in pregnancy test kits where latex particles are coated with human chorionic gonadotropin. Anti-HCG is added to the test urine sample on the slide. After mixing, the latex particles coated with human HCG are added. If there is agglutination and clumping of latex particles it indicates the samples negative. If the agglutination does not take place then the sample is positive for pregnancy.

4.7 ISOLATION OF PURE ANTIBODIES

There is often a need to isolate pure antibodies, which may be either antigen-specific or non-specific immunoglobulin. Isolation of non-specific immunoglobulin from serum is done by sequential protein fractionation steps, which may include:

1. *Precipitation of the gamma globulins* in 30–50% ammonium sulphate.
2. *Gel filtration* to obtain molecules of the correct size.
3. *Ion exchange chromatography* to isolate molecules, which are charged at neutral pH.

Isolation of antigen-specific immunoglobulin is carried out by **affinity chromatography** using antigen coupled to Sepharose. This technique is used where the isolation of pure antibodies or pure antigen is the objective.

Another way of obtaining pure antibody of a defined specificity is to produce **monoclonal antibodies** from cells in culture. An immortal clone of cells that manufactures a single antibody of defined specificity is created, by fusing myeloma cells with normal plasma cells that are actively producing specific antibody. Production can be maintained indefinitely through this clone.

4.8 RADIOIMMUNOASSAY

A very sensitive technique for detecting antigen or antibody is **radioimmunoassay (RIA)**, developed by Berson and Yallow in 1960. The principle of RIA involves competitive binding of radiolabelled antigen and unlabelled antigen to a high-affinity antibody. A high concentration of labelled antigen is mixed with the antibody, so that all antigen-binding sites of the antibody are saturated. Then increasing amounts of test sample, containing unlabelled antigen of unknown concentration are added. The two kinds of antigen compete for the available binding sites on the antibody, as the antibody does not distinguish labelled from unlabelled antigen. With increasing concentrations of unlabelled antigen, more and more labelled antigen is displaced from the binding sites. Even a small amount of unlabelled antigen added to the assay mixture causes a decrease in the amount of radioactive antigen bound, and this decrease is proportional to the amount of unlabelled antigen added. By measuring the free labelled antigen in the solution, it is possible to determine the concentration of unlabelled antigen.

To find the amount of labelled antigen bound, the Ag-Ab complex is precipitated to separate it from free unbound antigen, and the radioactivity of the precipitate is measured. A standard curve is developed using unlabelled antigen samples of known concentrations, and from this plot the amount of antigen in the test mixture is precisely determined (Figure 4.5). The antigen is generally labelled with ^{125}I and ^{3}H.

This is used for quantitation of hormones such as thyroxine and insulin.

4.9 ENZYME-LINKED IMMUNOSORBENT ASSAY (ELISA)

Enzyme-Linked Immunosorbent Assay (ELISA) is similar in principle to RIA, but uses an enzyme instead of a radioactive label. This method approaches the sensitivity of RIA, and is safer and less costly. An enzyme-antibody conjugate reacts with a colourless substrate to generate a coloured reaction product. The intensity of colour is measured by taking the optical density on a colorimeter. The commonly used enzymes and their colour substrate complexes are:

1. Horse radish peroxidases(enzyme)-substrate: Hydrogen peroxide + Orthophenylene diamine.
2. Alkaline phosphatase (enzyme)-substrate: p-nitrophenyl phosphate.
3. β-galactosidase (enzyme)-substrate: o-nitrophenyl-β-d-galactopyranoside.

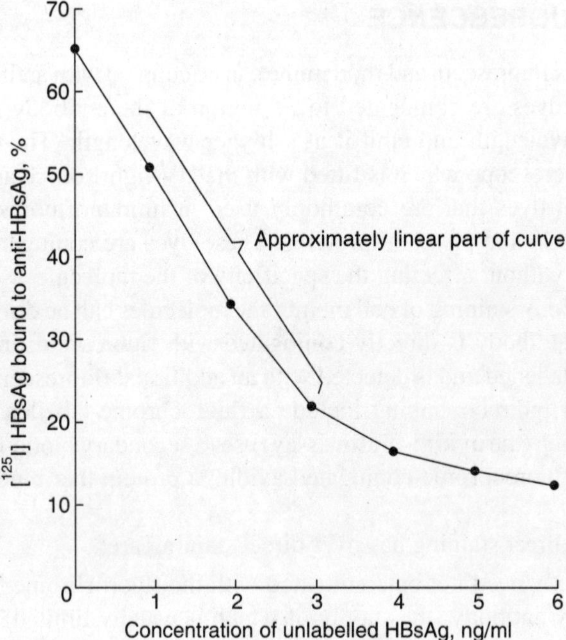

Figure 4.5 A solid-phase radioimmunoassay to detect hepatitis-B virus in blood samples.

A number of variations of this assay have been developed for qualitative and quantitative measurement of antigen and antibody. They are indirect ELISA, sandwich ELISA, competitive ELISA, chemiluminescence and ELISpot assay.

Enzyme-Linked Immunosorbent Spot assay (ELISpot assay)

This is a modification of the hemolytic plaque assay. Here the plasma cells are conjugated without using sheep RBC. The antigen primed lymphocytes are incubated in a petridish which is coated with antigen. As the plasma cells secrete antibody, it binds to the antigen in the immediate vicinity of the plasma cells. Excess lymphocytes are removed after sometime and the petridish is washed with saline. ELISA then visualizes the bound antibody. Enzyme substrate reaction leads to coloured spot formation, which are counted for enumeration of plasma cells.

Western blotting

Western blotting is used for identification of a specific protein in a complex mixture of proteins. A protein mixture is separated by SDS-polyacrylamide gel electrophoresis, and the protein bands are transferred to a nitrocellulose membrane by electrophoresis. The nitrocellulose membrane is flooded with radiolabelled or enzyme-linked antibody that is specific for the protein of interest, and the individual protein bands are identified. The Ag-Ab complexes formed on the band containing the protein of interest and recognized by the antibody can be visualized in a number of ways. This method is widely used for the confirmatory testing for HIV.

4.10 IMMUNOFLUORESCENCE

Fluorescent dyes such as fluorescein and rhodamine can be coupled with antibody without destroying their specificity. Both dyes are conjugated to Fc region of the antibody molecules. These dyes absorb light at one wavelength and emit it at a longer wavelength. The emited light is viewed through fluorescent microscope which is fitted with the UV light source and excitation filter.

Some fluorescent dyes that are commonly used in immunofluorescence are fluorescein, rhodamine, phycoerythrin and phycobiliproteins. These dyes are conjugated to the Fc portion of the antibody molecule without affecting the specificity of the molecule.

Fluorescent-antibody staining of cell membrane molecules can be direct or indirect. In **direct staining**, the specific antibody is directly conjugated with fluorescein; in **indirect staining** the primary antibody is unlabelled and is detected with an additional fluorescein-labelled reagent. The reagents developed for indirect staining include a fluorochrome-labelled anti-isotype antibody. Another indirect approach (the **avidin-biotin** assay) uses a secondary biotin-conjugated anti-isotype antibody followed by fluorochrome-conjugated avidin, a protein that binds with high affinity to biotin.

The advantages that indirect staining has over direct staining are:

- Primary antibody need not be conjugated with the fluorochrome. There is, therefore, no loss of primary antibody (the supply of which is usually limited) that occurs during the conjugation reaction.
- There is an increased sensitivity of staining as multiple molecules of the fluorochrome reagent bind to each primary antibody molecule, thereby increasing the amount of light emitted.
- The secondary antibody that is labelled can be used in a number of different reactions as it is anti-isotype antibody.

Fluorescence quenching, fluorescence polarization and fluorescence enhancement

These are some other techniques that employ fluorescence labelled molecules to detect interaction between antigen and antibody.

Flow cytometry and fluorescence

Flow cytometry provides quantitative data about antigen-antibody reactions. One version of flow cytometry is called FACS (Fluorescence-Activated Cell Sorter), which automates the analysis and separation of cells labelled with fluorescent antibody. The instrument counts each cell as it passes a laser beam and records the level of fluorescence the cell emits. Use of the instrument for finding the number in a cell population is called analysis, while its use to place the cells having different patterns of reactivity into different containers is called cell sorting.

A common use of FACS is to determine the kind and number of white blood cells in each population in patients' blood samples. One can obtain the following information:

- The number of cells that express a target antigen, and also as a percentage of cells passing the beam.
- The distribution of cells in a sample population according to antigen densities as determined by fluorescence intensity.
- The size of cells, which is derived from analyzing light-scattering properties of different cells in a population.

Laser Nephelometry: Nephelometry is the measurement of light scattering by particles in suspension. When optically clear solutions of antigen and antibody are mixed, the resulting precipitate makes the mixture appear cloudy. The quantity of immune complexes formed can be measured by shining the beam of a helium-neon laser through the solution. This technique can be used for a rapid antibody assay.

MULTIPLE-CHOICE QUESTIONS

1. Which of the following is a quantitative precipitation technique?
   (a) gel diffusion (b) western blot
   (c) radial immunodiffusion (d) immunoelectrophoresis
2. The most sensitive immunological test in terms of the amount of antibody detectable is:
   (a) complement fixation test (b) radioimmunoassay
   (c) gel precipitation test (d) agglutination test
3. The ligand for the small protein avidin is:
   (a) biotin (b) ferritin
   (c) fluorescein (d) antiglobulin
4. Immune complex precipitates are formed in:
   (a) antigen access zone (b) antibody access zone
   (c) the zone of equivalence (d) in the absence of an electrolyte
5. The antiglobulins are:
   (a) incomplete antibodies (b) agglutinating antibodies
   (c) antibodies against immunoglobulins (d) complement fixing antibodies
6. Cross-reactions may be due to:
   (a) non-specific antibodies (b) dissimilar epitopes on antigens
   (c) similar epitopes on antigens (d) chemical reactions with antigens
7. The ability of an antibody that binds blood group antigens to also react with some *E. coli* bacteria is an example of:
   (a) cross-reactivity (b) autoimmunity
   (c) passive immunity (d) transfusion reaction

REVIEW QUESTIONS

4.1 Give an outline of the procedures for direct and indirect fluorescent antibody tests.

4.2 Write notes on:
(i) Radioimmunoassay
(ii) Enzyme linked immunosorbent assay
(iii) Antigen antibody cross-reactivity
(iv) Mechanism of antigen antibody interaction.

5

Generation, Activation and Differentiation of Lymphocytes

5.1 T-CELL MATURATION, ACTIVATION AND DIFFERENTIATION

The maturation of progenitor T-cells in the thymus and the activation of mature T-cells in the periphery are influenced by the MHC molecules. MHC restriction distinguishes the antigen recognition ability of T-cell from that of B-cell. T-cells undergo maturation by a selection process and during maturation only MHC restricted and non-self reactive T-cells undergo maturation. Those T-cells restricted by Class I MHC develop into $CD8^+$ T-cells, and those restricted by Class II MHC give rise to $CD4^+$ T-cells.

The progenitors of T-cells migrate from the bone marrow into the thymus under the influence of chemotactic factors that are secreted by the thymic epithelial cells. This process starts on about the 11th day of gestation in mice, and in the 8th or nine week of gestation in humans. The generation of a mature T-cell population with a diverse TCR repertoire is divided into an early phase in which the development of thymocytes can proceed in the absence of mature TCR, and a late phase in which further maturation critically depends on the cells expression of a functional $\alpha\beta$ TCR. The early phase covers all the developmental stages prior to the appearance of $CD4^+CD8^+$ positive thymocytes. T-cell maturation occurs in the thymus when progenitor T-cells from the bone marrow enter the thymus and rearrange their TCR genes.

On arrival in the thymus, progenitor T-cells do not express surface molecules characteristic of T-cells. They lack detectable CD4 and CD8, and are therefore called **Double Negative (DN)** cells. The expression of other cell surface molecules particularly C-Kit, CD44 and CD25 marks the developmental progression of the DN population in the early phase.

The most immature thymocytes lack TCR CD3 expression and reside in the $CD4^-$ $CD8^-$ populations which are called Triple Negatives (TNs). This population is only about 1 to 2 per cent of all thymocytes and passes through a series of differentiation stages. These stages ultimately lead to the coexpression of CD4 and CD8 and the generation of Double Positive (DP) thymocytes.

Pre-T-cell receptor (Pre-TCR) is expressed on thymocytes that lack CD4 and CD8 markers on the surface. This receptor consists of a CD3 protein and a disulphide linked heterodimer made of β chain of the TCR and a 33 kD type I transmembrane glycoprotein which is called pre-Tα(pTα) covalently associated with TCRβ. It belongs to the immunoglobulin superfamily and is encoded by a non-rearranging gene.

In the final stages of maturation two different development pathways are followed, which generate functionally distinct $CD4^+$ and $CD8^+$ sub-populations that exhibit Class II and Class I MHC restriction, respectively. T-cell maturation involves rearrangements of the germ-line T-Cell Receptor (TCR) genes and the expression of various membrane markers. In the thymus, developing T-cells, known as **thymocytes**, proliferate and differentiate along different developmental pathways, which generate distinct sub-populations of mature T-cells. The T-cells diversify by a pair of selection processes.

5.1.1 Positive Selection

Positive selection is an interaction between immature double positive thymocytes with thymic epithelial cells. This interaction is through their TCR which establishes the contact with Class I and Class II MHC expressed on thymic epithelial cells. Those thymocytes which fail to establish such contacts undergo apoptosis. Only those thymocytes whose $\alpha\beta$ TCR heterodimer recognizes a self-MHC molecule are selected for survival. **Positive selection** permits the survival of only those T-cells whose TCRs recognize self-MHC molecules.

5.1.2 Negative Selection

The **negative selection** eliminates T-cells that react too strongly with self-MHC. The MHC restricted population of positively selected thymocytes shows two types of TCR affinity (high and low) to self-MHC. Those thymocytes which show high affinity with self-MHC molecules undergo negative selection by interaction with bone marrow derived dendritic cells or macrophages in the medulla of the thymus. Tolerance to self antigen is achieved by eliminating self reactive T-cells and only allowing maturation of T-cells that are specific for foreign antigens and altered self molecules. These processes generate a primary T-cell repertoire that is self-tolerant (Table 5.1).

Table 5.1 Characteristics of T-cell Selection in the Thymus

Property	*Positive Selection*	*Negative Selection*
Site	Cortex	Medulla
Stromal cells involved	Epithelial cells	Macrophages and dendritic cells
Selection mechanism	Survival of thymocytes bearing receptors for self-MHC	Elimination of thymocytes bearing high-affinity receptors for self-MHC alone or self-antigen + self-MHC
Immune consequence	Self-MHC restriction	Self-tolerance

The thymocytes that undergo productive TCR gene rearrangement are not put through the selection process which involves positive selection or negative selection. The positive selection ensures $\alpha\beta$ TCR expressed on the T-cell of an individual will bind to self-MHC. The cells which fail to undergo positive selection are eliminated within the thymus. Those thymocytes which show high affinity receptors for cells MHC molecules undergo death by apoptosis. The processes are necessary to generate mature T-cells that are MHC restricted and self-tolerant.

About 98 per cent of all thymocytes do not mature. They die by apoptosis within the thymus. The earliest thymocytes that lack CD4 and CD8 differentiate along two pathways. Those thymocytes that make a productive rearrangement of the $\gamma\delta$ TCR genes develop into $CD4^-$, $CD8^-$, $CD3^+$ $\gamma\delta$ T-cells, which are in a minority. The majority of double negative thymocytes rearrange the $\alpha\beta$ TCR genes and develop into either $CD4^+$, $CD3^+$ $\alpha\beta$ T-cells or $CD8^+$, $CD3^+$ $\alpha\beta$ T-cells.

Most double negative thymocytes progress down a different development pathway. They stop proliferating and begin to rearrange the TCR β-chain genes and express the β-chain. These β-chains combine with a 33 kDa glycoprotein known as pre-Tα chain and associate with the CD3 group to form a complex called **pre-T-Cell Receptor (pre-TCR)**. Immature thymocytes in the $\alpha\beta$ pathway express a pre-T-cell receptor.

The pre-TCR recognizes some intra-thymic ligand and transmits the signal through the CD3 complex that activates a protein tyrosine kinase. Once a signal has been transmitted through the pre-TCR, it halts further β-chain gene rearrangement and induces expression of both CD4 and CD8. The thymocytes are now called **Double Positive (DP)**, or $CD4^+8^+$ cells, which begin to proliferate.

Double positive thymocytes that express the $\alpha\beta$ TCR-CD3 complex and survive thymic selection develop into either **single-positive $CD4^+$** thymocytes or **single-positive $CD8^+$** thymocytes. Thus, in the thymus there is:

- Positive selection of thymocytes bearing receptors capable of binding self-MHC molecules, which results in **MHC-restriction**.
- Negative selection by elimination of thymocytes bearing high affinity receptors for self-MHC molecules alone or self-antigen presented by self-MHC, which results in **self-tolerance**.

Both these processes are necessary to generate mature T-cells that are self-MHC restricted and self-tolerant. Thymic stromal cells, including epithelial cells, macrophages, and dendritic cells play essential roles in positive and negative selection. During positive selection only those cells whose $\alpha\beta$ TCR heterodimer recognizes a self-MHC molecule are selected for survival.

5.2 T_h-CELL ACTIVATION

The central event in the generation of both humoral and cell mediated immune responses is the activation and the clonal expansion of T_h-cells. The activated T_h-cell progresses through the cell cycle, proliferating and differentiating into memory cells or effector cells. After T_h-cells interact with antigen, numerous genes are activated.

Superantigens are viral or bacterial proteins that bind simultaneously to the Vβ domain of a T-cell receptor and to the α chain of a Class II MHC molecule. Crosslinking of a T-cell receptor and the Class II MHC molecule by a superantigen produces an activating signal that induces T-cell activation and proliferation. Activation of T_h-cell is initiated by an antigen presenting cell which presents antigenic peptide on its surface in the groove of a Class II MHC molecule.

T_h-cell activation is initiated by interaction of the TCR-CD3 complex with a peptide-MHC complex on an antigen-presenting cell. The activating signal is transduced by the TCR-CD3 complex and regulated by co-receptors CD4, CD8 and CD5. Signal tranduction is accomplished by a series of protein phosphorylation and dephosphorylation events catalyzed by protein kinases and protein phosphatases.

In addition to the signals mediated by T-cell receptor and its associated accessory molecules (Signal 1), activation of the T_h-cell requires a co-stimulatory signal (Signal 2) provided by the antigen-presenting cells.

T_h-cell recognition of an antigenic peptide-MHC complex on an antigen presenting cell results either in activation and clonal expansion or in a state of non-responsiveness called clonal anergy. The presence or absence of the co-stimulatory signal (Signal 2) determines whether activation results in clonal expansion or clonal anergy.

Only dendritic cells, macrophages, and B-cells present antigen together with Class II MHC molecules and deliver the co-stimulatory signal necessary for complete T-cell activation that leads to proliferation and differentiation. The antigen-presenting cells differ in their ability to display antigen and to deliver the co-stimulatory signal.

Class II MHC-peptide complex on APC establish the contact with TCR-CD3 complex of T_h-cell resulting in activating signals. There is a cascade of biochemical events that induces resting T_h-cell to enter the cell cycle G0 to G1, which results in the expression of high affinity autocrine receptor for IL-2. IL-2 is the T-cell growth factor; it helps in T_h-cell proliferation, progression and differentiation into effector and memory T-cells. Interaction of T_h-cell with antigen on Class II MHC-APC will lead to activation of many genes. The gene activation occurs at different intervals. There are immediate genes that are activated within half an hour of antigen recognition, for example, c-Fox, c-Myc, NFAT. Early genes are those which are expressed within one to 2 hours of antigen recognition for example, IL-2, IL-2R, IL-3, IFN-γ. Late genes are expressed more than two days after antigen recognition and are responsible for secretion of various adhesion molecules.

5.3 T-CELL DIFFERENTIATION

Naive T-cells survive for only about five to seven weeks in the absence of antigen stimulation. Each naive T-cell recirculates from the blood to the lymph nodes and back again every 12 to 24 hours.

$CD4^+$ and $CD8^+$ T-cells leave the thymus and enter the circulation as resting cells in the G0 stage of the cell cycle as naive T-cells, which have not yet encountered antigen. They continually recirculate between the blood and lymph systems. During recirculation, naive T-cells reside in secondary lymphoid tissues such as lymph nodes. If a naive cell does not encounter antigen in a lymph node, it exits through the efferent lymphatics, and drain into the thoracic duct to rejoin the blood.

On the other hand, if a naive T-cell recognizes an antigen-MHC complex on an appropriate antigen-presenting cell or target cell it gets activated, initiating a *primary response.* After 48 hours of activation, the naive T-cell enlarges into a blast cell and begins undergoing repeated rounds of cell division. Blast cells divide two to three times per day and keep on doing so for four to five days thereby generating a large population of clones which differentiate into memory or effector T-cells.

Activation depends on the signal induced by engagement of the TCR complex and a co-stimulatory signal induced by the CD28-B7 interaction. These signals trigger entry of the T-cell into the G1 phase of the cell cycle and induces transcription of the gene for IL-2 and the α chain of the high affinity IL-2 receptor. It also increases the half-life of IL-2 mRNA. Secretion of IL-2 and its subsequent binding to the high affinity IL-2 receptor induces the activated naive T-cells to proliferate and differentiate. T-cells activated in this way divide 2–3 times per day for 4–5 days, generating a large clone of progeny cells which differentiate into memory or effector T-cell populations. Effector cells are derived from both naive and memory cells after antigen activation. The effector T-cells have a short lifespan ranging from few days to a few weeks. They differ from the naive T-cells in respect to cell surface markers which contribute to their recirculation ability. One subset of CD4 effector cells is called T_h1 subset which secretes IL-2, IFN-γ and TNF-β and the cells are responsible for the activation of cytotoxic T-cells and are also involved in the delayed type of hypersensitivity. The other subset T_h2 secretes IL-4, IL-5, IL-6 and IL-10 and serves as effector helper cell for B-cell activation.

5.4 MEMORY T-CELLS

The **memory T-cell** population is derived from both naive T-cells and from effector cells after they have encountered antigen. Memory T-cells are generally long-lived, antigen-activated T-cells that show heightened response to the subsequent challenge with the same antigen, giving rise to a *secondary response*. Memory T-cells are responsible for generating secondary response and like the naive T-cells they are resting cells with cell division activity arrested at G0 stage, but they have less stringent requirements for the activation than naive T-cells.

Memory T-cells are activated by macrophages, dendritic cells and T-cells, whereas the naive T-cells are activated only by dendritic cells. Memory T-cells express a wide variety of cell adhesion molecules which enhances their ability of interaction with APCs.

5.5 B-CELL GENERATION, ACTIVATION AND DIFFERENTIATION

Development of B-cells from committed progenitors to terminally differentiated plasma cells and memory B-cells is the multistep process which can be divided into three broad stages: generation of mature, immunocompetent B-cells(maturation), activation of mature B-cells by interaction with antigen, and differentiation of activated B-cells into plasma cells and memory cells.

The bone marrow generates B-cells in an orderly sequence of Ig-gene rearrangements, which proceeds in the absence of antigen. Maturation of the cells occurs in the bone marrow microenvironment. This stage is called antigen-independent phase. Immature B-cells having membrane-bound immunoglobulins, mIgM and mIgD, leave the bone marrow and migrate to the periphery to become resting mature B-cells or naive B-cells. These naive cells have not encountered the antigen and circulate in the blood and lymph and are carried to the secondary lymphoid organs like the spleen and lymph nodes. This is known as the antigen-independent phase of B-cell development.

A mature B-cell leaves the bone marrow expressing membrane-bound immunoglobulin (mIgM and mIgD) with a single antigenic specificity. These **naive B-cells**, which have not encountered antigen, circulate in the blood and lymph and are carried to the secondary lymphoid organs, notably the spleen and the lymph nodes. If a B-cell encounters an antigen which binds to its membrane-bound immunoglobulins then the cell will undergo activation. Activation of B-cells leads to proliferation and differentiation that result in the formation of antibody secreting plasma cells and memory B-cells. Antigen committed B-cells undergo affinity maturation where class switching of immunoglobulin gene occurs.

5.5.1 Affinity Maturation

Affinity maturation is the progressive increase in the average affinity of the antibodies produced while **class switching** is the change in the isotype of the antibody produced by the B-cell from μ to γ, α, or ε. Since B-cell activation and differentiation in the periphery requires antigen, this stage comprises the antigen dependent phase of B-cell development. Throughout the lifespan of an animal, only a small fraction of the possible antibody diversity is ever generated.

5.5.2 B-cell Maturation

The generation of mature B-cells first occurs in the embryo and continues throughout life. Before birth, the yolk sac, foetal liver, and foetal bone marrow are the major sites of B-cell maturation; after birth the generation of mature B-cells occurs in the bone marrow.

The development of B-cells begins from the lymphoid stem cells which differentiate into B-cell lineage called progenitor B-cell (pro-B-cell). B-cell development begins as lymphoid stem cells differentiate into the **progenitor B-cells (pro-B-cell),** which proliferate within the bone marrow. The cells committed to B-cell lineage have the marker CD45R and a transmembrane tyrosine phosphatase. Pro-B-cells also express CD19, CD43 and CD24 on the surface. The display of the pre-B-cell receptor (pre-BCR) is characteristic of the pre-B-cell stage. Proliferation and differentiation of the pro-B-cells into **precursor-B-cells (pre-B-cells)** requires the microenvironment provided by the bone marrow stromal cells. Stromal cells secrete various cytokines like IL-7. They also interact with the pro-B- and pre-B-cells. This interaction is very necessary for the development of B-cells. Cell adhesion molecules like VLA-4 on pro-B-cells and VCAM-1 on stromal cells help in the interaction. After this interaction is established, other molecules like c-Kit on the pro-B-cells and Stem Cells Factor (SCF) on the stromal cells establish contact and induce tyrosine kinase activity. The pro-B-cells start dividing and differentiate into pre-B-cells expressing a receptor for IL-7. Stromal cells secrete IL-7 which binds to pre-B-cell receptors and downregulates the adhesion molecule contact so as to release the pre-B-cells from the stromal cell contact. The detached pre-B-cells continue to proliferate with the help of IL-7 stimulation and undergo maturation.

Development and maturation of B-cells depends on Ig gene rearrangements. In the pro-B-cell stage the first rearrangement to occur is the heavy chain D_H to J_H rearrangement. This is followed by a V_H to D_HJ_H rearrangement. Upon completion of heavy chain rearrangement, the cell is classified as the pre-B-cell.

Development of pre-B-cell into immature B-cell requires a productive light chain gene rearrangements. After rearrangement of the light chain the cells are known as immature B-cells, and lose the pre-BCR and no longer express CD25. Only one light gene isotype is expressed on the membrane of a B-cell. This rearrangement commits the now immature B-cell to a particular antigenic specificity determined by the cell's heavy chain VDJ sequence and the light chain VJ sequence. The immature B-cell express mIgM on the cell surface.

Transition of cells from pre-B- to immature B-cell stage is accompanied by productive VJ gene rearrangements that allow expression of conventional light chains. At pro-B-cell stage recombinase enzymes RAG-1 and RAG-2 appear together with terminal deoxynucleotidyl transferase (TdT), which play an important role in heavy chain, light chain gene rearrangements. Further development of B-cells leads to the formation of mature B-cells with fully expressed mIgM and mIgD on their membrane. These cells are then exported from the bone marrow to peripheral lymphoid organs.

The developmental process that produces naive mature B-cells is completed when the IgD class of Ig joins IgM on the cell surface.

During B-cell maturation, sequential Ig-gene rearrangement transforms the pro-B-cell into an immature B-cell, which expresses mIgM with a single antigenic specificity. A pre-B-cell receptor is required for maturation to continue to the immature B-cell stage. Further development, including changes in RNA processing, yields mature B-cells, that express both IgM and IgD.

The bone marrow produces about 5×10^7 cells/day, but only about10 per cent are actually recruited into the recirculating B-cell pool. These immature B-cells then migrate to the spleen where they have a half-life of about four days. Here they differentiate into mature B-cells with a lifespan of about 15 weeks. About 90 per cent of the immature B-cells produced in the bone marrow are lost, a majority of which are the auto reactive B-cells.

This is because of a selection procedure that operates once the cell reaches the immature stage with high IgM and low IgD expression. This loss is called negative selection. In the bone marrow itself, the B-cells that express autoantibodies (mIgM) against self antigens are also subjected to clonal deletion. Cross-linking of IgM on immature B-cells leads to cell death. It has been experimentally established that immature B-cells with monoclonal antibody against constant region cross-links with mIgM and results in death of immature B-cells. A similar process occurs in the bone marrow when immature B-cells encounter the cells antigens.

This is because of **negative selection** and subsequent elimination (**clonal deletion**) of immature B-cells that express autoantibodies against self-antigens in the bone marrow.

5.5.3 B-cell Activation and Proliferation

Naive B-cells are non-dividing B-cells in the G0 stage of the cell cycle. Activation drives the resting cell into the cell cycle, progressing through G1 into the S phase, in which DNA is replicated. Once the cell has reached S phase, it completes the cell cycle, moving through G2 and into mitosis.

After export of mature B-cells from the bone marrow, activation, proliferation and differentiation occur in the periphery and requires antigen. Activating signals fall into two categories, namely, the competence and progression signals. **Competence signals** drive the B-cell from G0 into early G1. **Progression signals** then drives the cell from G1 into S and ultimately to cell division and differentiation.

Antigen driven activation of B-cells goes through two different methods on the basis of antigen involved. Depending on the nature of the antigen, B-cell activation proceeds by different routes that are either dependent upon T_h-cells or independent of them. The B-cell response to **Thymus Dependent (TD) antigens** requires direct contact with T_h-cells.

Antigens that can activate the B-cells in the absence of direct participation by T_h-cells are known as **Thymus Independent (TI) antigens** which are of TI-1 and TI-2 types.

The TI antigens are multivalent and therefore, induce strong stimulation of B-cells by cross-linking mIg molecules, which act as signal 1. Type 1 antigens are bacterial cell wall components including lipopolysaccharides (LPS). In addition type 1 TI antigens also contain an additional component that provides Signal 2. Most TI-1 antigens are polyclonal B-cell activators, and at high concentrations stimulate proliferation and antibody secretion by as many as one third of all B-cells. The prototypic TI-1 antigen is lipopolysaccharide. TI-1 are not B-cell mitogens and therefore, do not act as polyclonal activators. Type 2 TI antigens are bacterial cell wall polysaccharides with repetitive units conjugated to polymeric proteins, for example bacterial flagellins. TI-1 antigens activate both mature and immature B-cells.

Type 2 TI antigens show extensive cross-linking with mIg receptor, generates an extensive competence signal, which itself leads to progression signal that leads to progression of B-cells. TI-2 antigens activate B-cells by extensively cross-linking the mIg receptor. TI-2 antigens activate only the mature cells, and inactivate immature B-cells. Although B-cell response to TI-2 antigens do not require direct involvement of T_h-cells, cytokines derived from T_h-cells are required for efficient B-cell proliferation and for class-switching isotypes other than IgM. Type 1 TI antigen seems to need an additional signal in the form of cytokine to enter into progression stage.

The response to TI antigens is generally weaker, no memory cells are formed, and IgM is the predominant antibody that is secreted. This reflects a low level of class switching. Thus T_h-cells have an important role in generating memory cells, affinity maturation, and class switching to other isotypes.

The binding of the Ig receptor by antigen activates intracellular signalling pathways.

Antigen-driven activation and clonal selection of naive B-cells leads to generation of plasma cells and memory B-cells. In absence of antigen-induced activation, naive B-cells in the periphery have a short lifespan, dying within a few weeks by apoptosis.

During B-cell differentiation affinity maturation, somatic hyper mutation and class switching occurs. This is followed by the centroblasts differentiating into memory B-cells and plasma cells. A CD40 ligand represents a key negative signal that prevents centroblasts from entering into terminal plasma cell differentiation. As long as CD40 ligand is provided centroblasts remain as it is in the presence of IL–2 and IL–10. Withdrawal of this ligand results in the centroblasts rapidly differentiating into plasma cells.

The long-lived plasma cells are terminally differentiated B-cells that contribute to immune memory through the continual production of high affinity antibody. Memory B-cells generally express isotype switched and affinity matured membrane Ig, and are required to cognate memory T_h-cell regulation for the response to antigen.

Once the antigen binds to mIg receptor on B-cells, the antigen is processed into peptides. Binding of antigen to B-cell receptor mIg results in signal transduction leading to a sequence of events in the cytoplasm and gene action to upregulate the production of certain membrane associated molecules such as Class II MHC, costimulatory molecule B-7 and the receptors for the growth factors and cytokines. Enhanced expression of B-7 and Class II MHC molecule on B-cells increases its ability to function as an antigen presenting cell for T_h-cell activation. T–B-cell conjugated is formed only when B-cell presents the processed antigenic peptide on its Class II MHC molecule to T_h-cell through TCR. This contact is essential for the directional release of cytokines by the T_h-cells and also for B-cell activation. Three T_h-cell derived cytokines, IL-2, IL-4 and IL-5 provide progression signal for B-cells to proliferate. Proliferation of B-cells leads to three different events: formation of plasma cells, memory B-cells, class switching and affinity maturation. These events require signals provided by T_h-cells or follicular dendritic cells. This event in the B-cell differentiation is the result of somatic hyper mutation and antigen selection of high affinity clones.

5.6 THE HUMORAL RESPONSE

Activation of naive B-lymphocytes elicits a *primary response*, while activation of memory lymphocytes leads to a *secondary response*. In both cases activation leads to production of secreted antibodies of various isotypes, which differ in their ability to mediate specific effector functions.

A primary response to the antigen is characterized by a lag phase, during which naive B-cells undergo clonal selection, subsequent clonal expansion, and differentiation into memory cells or plasma cells. The lag phase is followed by a logarithmic increase in serum antibody level, which reaches a peak, plateaus for a variable time, and then declines. Eight or nine successive cell divisions of activated B-cells generate plasma and memory cells. During a primary humoral response, IgM is secreted initially, followed by a switch to an increasing proportion of IgG. It is thus a biphasic response. A primary response can last for various periods, depending on the persistence of the antigen. The memory B-cells formed during a primary response have variable lifespans, and some persist for the life of the individual.

The secondary response has a shorter lag period, is more rapid, reaches a greater magnitude, and lasts longer. This is also characterized by secretion of antibody with higher affinity for the antigen, and isotypes other than IgM are formed. The population of memory B-cells specific for a given antigen is considerably larger than the population of corresponding naive B-cells. Moreover, the memory B-cells are more easily activated than naive B cells are. Affinity maturation and class switching are responsible for the higher affinity and different isotypes exhibited in a secondary response (Tables 5.2 and 5.3, Figure 5.1).

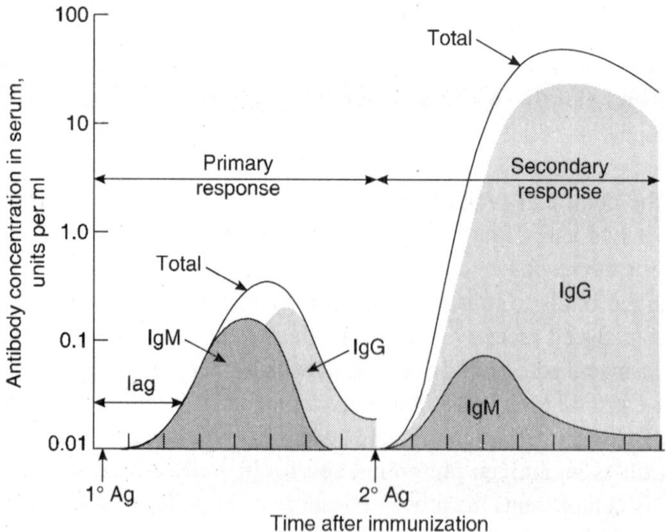

Figure 5.1 Concentration and isotype of serum antibody following primary and secondary immunization with antigen.

Affinity maturation, class switching, and formation of plasma and memory cells all take place during B-cell proliferation and differentiation within germinal centres.

Table 5.2 Comparison of Primary and Secondary Immune Responses

Properties	*Primary Response*	*Secondary Response*
Responding B-cell	Naive (virgin) B-cell	Memory B-cell
Lag period following antigen administration	Generally 4–7 days	Generally 1–3 days
Time of peak response	7–10 days	3–5 days
Magnitude of peak antibody response	Varies depending on antigen	Generally 100–1000 times higher than primary response
Isotype produced	IgM predominates early in the response	IgG predominates
Antigens	Thymus-dependent and thymus-independent	Thymus-dependent
Antibody affinity	Lower	Higher

Table 5.3 Comparison of Naive and Memory B-cells

Properties	*Naive B-cell*	*Memory B-cell*
Membrane markers		
Immunoglobulin	IgM, IgD	IgM, IgD, IgG, IgA, IgE
Complement receptor	Low	High
Anatomic location	Spleen	Bone marrow, lymph node, spleen
Lifespan	Short-lived	May be long-lived
Recirculation	Yes	Yes
Receptor affinity	Lower average affinity	Higher average affinity
Adhesion molecules	Low I CAM-1	High I CAM-1

5.6.1 Generation of Plasma Cells and Memory B-cells

After selection of lymphocytes bearing high affinity mIg for antigen displayed on follicular dendritic cells, the lymphocytes differentiate into plasmablasts and memory B-cells. Plasma cells do not have any membrane-bound immunoglobulins, and instead synthesize high levels of secreted antibody. Differentiation of mature B-cells into plasma cells requires a change in RNA processing so that the secreted form of the heavy chain rather than the membrane form is synthesized. The rate of transcription of heavy and light chains is greater in plasma cells than in the less differentiated B-cells.

Development of B-cells involves antigen-independent maturation in the bone marrow, and antigen-dependent activation and differentiation of mature B-cells in the periphery, resulting in antibody-secreting plasma cells and memory B-cells.

Induction of immune responses by self-antigens can have serious consequences. When a self-antigen is expressed in the bone marrow, negative selection of self-reactive immature B-cells occurs. This process ensures that an individual does not mount autoimmune responses.

MULTIPLE-CHOICE QUESTIONS

1. What is the major activity occurring in germinal centres?
   (a) antibody production (b) antigen trapping
   (c) somatic mutation (d) antigen presentation
2. The central paradigm of the immune response is called:
   (a) somatic mutation (b) antigen presentation
   (c) MHC restriction (d) clonal selection
3. The need for a T_h-cell and a B-cell to recognize two different epitopes on an antigen is called:
   (a) linked recognition (b) MHC restriction
   (c) Antigenic variation (d) covalent bonding.

REVIEW QUESTIONS

5.1 Give an account of the following:
(i) T-cell maturation
(ii) T-cell activation
(iii) T-cell differentiation

5.2 Give an account of the following:
(i) B-cell maturation
(ii) B-cell activation
(iii) B-cell differentiation

5.3 What is clonal selection? How does it occur?

5.4 Write a note on the generation of plasma cells and memory cells.

5.5 What is the advantage of possessing T helper-cells?

5.6 What is the biological basis of MHC restriction between antigen-presenting cells and helper T-cells?

Major Histocompatibility Complex

6.1 MAJOR HISTOCOMPATIBILITY COMPLEX (MHC)

Antigen recognition by T-lymphocytes is central to the generation and regulation of an effective immune response. The T-Cell antigen Receptor (TCR) is a disulphide-linked heterodimeric (either $\alpha\beta$ or $\gamma\delta$) glycoprotein that enables T-cells to recognize a diverse array of antigens. It is associated at the cell surface with a complex of polypeptides known as CD3 (Table 6.1).

Table 6.1 Comparison of MHC Class I and Class II Structure

	Class I	*Class II*
Loci include	HLA-A, B, and C H-2-K, D, and L	HLA-D (DP, DQ, DR, DO, and DN) H-2-1 (IA and IE)
Detected by	Serology	Serology and mixed lymphocyte reaction*
Distribution	Most nucleated cell	B-cells, macrophages, and other dendritic cells
Function	Present antigen to cytotoxic T-cells	Present antigen to helper T-cells
Result	T-cell-mediated cytotoxicity	T-cell-mediated help

*A mixed lymphocyte reaction occurs when lymphocytes from two individuals with different MHC Class II molecules are cultured together. Each population stimulates the other to divide.

Antigen processing requires fragmentation of antigen molecules inside cells, and linkage of these fragments to appropriate antigen-presenting molecules. These antigen-presenting molecules are called **histocompatibility** or **MHC molecules**. They are specialized receptor glycoproteins inherited in a gene complex called the **Major Histocompatibility Complex (MHC).** The absolute need for antigens to be presented bound to MHC molecules is called **MHC restriction**.

The host's cells carry MHC molecules on the surface, which are associated with self/non-self recognition. In infected cells these MHC molecules bind to and displace small peptides that come from the parasites. The complexes of parasitic peptides and host MHC molecules form the antigens that can be recognized by antigen receptors on cytotoxic T-lymphocytes. In this way the killer cells can identify and kill infected cells selectively, sparing healthy cells. One function of the MHC peptide complexes is therefore to signal that a cell is infected.

MHC peptides complexes are also important in the regulation of immune responses.

The recognition of a MHC-peptide complex on the surface of a cell is the critical event in the initiation of cellular and humoral immune responses.

It has been observed that if a tissue is transplanted between genetically non-identical (allogeneic) individuals, it usually triggers an immune response causing rejection of the transplanted tissue (allograft). Graft rejection occurs when major histocompatibility antigens (generally called Class I MHC and Class II MHC molecules) of the recipient do not match the histocompatibility antigens of the donor tissue. This genetic dissimilarity is referred to as incompatibility between donor and recipient MHC antigens. MHC molecules show unique specificity in each individual and can even discriminate between self and non-self antigens.

The major histocompatibility complex is a collection of genes arrayed within a long continuous stretch of DNA on chromosome 6 in humans and these genes are organized into regions encoding three classes of molecules. These encoded proteins are associated with intercellular recognition and antigen presentation to T-lymphocytes. All mammals possess a major histocompatibility complex in their genome. The MHC, called the H-2 complex in mice and the HLA complex in humans, is organized into three regions based on the type of molecules encoded. Class I molecules are coded by the K and D regions in mice and the A, B, and C regions in humans. Class II molecules are encoded by the IA and IE regions in mice and the DP, DQ, and DR regions in humans. Class III molecules in both humans and mice are encoded in the region between Class I and II. Many alleles exist for each Class I and Class II MHC gene. The entire set of MHC alleles on a chromosome is referred to as its haplotype.

In humans the MHC is located on the short arm of chromosome 6 and contains about 3.5 megabases, while in mice it is on chromosome 17. It encodes two sets of highly polymorphic cell surface molecules; termed MHC Class I and MHC Class II. The $\alpha\beta$ TCR recognizes processed antigen as peptide fragments bound to MHC Class I or Class II molecules. Both MHC and peptide residues associate with the TCR.

6.1.1 Genomic Organization of the MHC

In humans, the MHC genes region consists of separate loci designated as HLA-A, HLA-B, HLA-C, HLA-D, and HLA-DR (-DR, -DP, -DQ). At each locus one of a number of alleles may be present, which codes for an MHC antigen with specificity unique to that particular allele. MHC antigens are also known as MHC gene products, Human Leukocyte Antigens (HLA), histocompatibility antigens or tissue compatibility antigens.

The inheritance of MHC genes follows simple Mendelian inheritance. Each individual has two half-sets of genes (one half from each parent), which are expressed equally on the cell surface of an offspring. The combination of alleles at each locus on a chromosome is inherited from each parent as a unit (haplotype).

6.1.2 Class I MHC Genes

Class I genes code for MHC molecules found on the surface of most nucleated cells (and on red cells in some species). The Class I genes can be divided into the highly polymorphic genes called Class Ia genes, and those that show very little polymorphism, called Class Ib genes.

The human Class I region contains three loci, called HLA-A, HLA-B and HLA-C. Each locus encodes the heavy chain of a classical MHC Class I antigen and the whole region extends over 1.5 million bases of DNA on chromosome 6 (Figure 6.1).

Class I MHC genes encode glycoproteins that are expressed on the surface of all nucleated cells and they present peptide antigens or altered self-cells or the antigens of intracellular parasite. Class I MHC molecule is necessary for presenting antigen to T_C-cells. The genes in the A, B and C regions of the HLA complex encode the α chain. HLA-A has 23 α chain alleles, HLA-B has 49 α chain alleles and HLA-C has 8 α chain alleles. Polymorphism is a result of variations in the amino acid sequences in $\alpha 1$ and $\alpha 2$ domains. The $\beta 2$-microglobulin is encoded by a different gene on a different chromosome. Class Ib is less polymorphic and expressed in lower levels than the Class Ia.

6.1.3 Class II MHC Genes

Class II MHC genes are located in the HLA-D region. Class II genes code for a different set of polymorphic MHC molecules found only on the surface of macrophages, dendritic cells, and B-cells, and they are responsible for presentation of processed antigens to T_h-cells for recognition and discrimination.

Both Class Ia and Class II MHC molecules act as receptors that bind antigen fragments and present antigens to lymphocytes. Therefore they effectively regulate the immune responses.

The Class II region also contains genes that encode proteins involved in antigen presentation that are not expressed at the cell surface. The MHC Class II region is called HLA-D. It is located on the centromeric side of Class I and III regions on chromosome 6. Within the HLA-D region are five loci, which are designated as DN, DO, DP, DQ, DR, and there are genes for proteins involved in transporting antigen fragments across the endoplasmic reticulum. The order of the five loci present is DP, DN, DO, DQ, DR. Within each locus are genes for the α chains that are designated as A and for the β chains which are designated as B. The DN locus consists of a single A gene while the DO locus has a single B gene. The protein products of these genes have never been found. At the DP locus are two A genes and two B genes; out of these four genes, the genes DPA2 and DPB2 are pseudogenes that cannot form a protein. Thus, this locus has one functional gene each for the α chain and the β chain. At DQ there are two A genes and three B genes. Out of these genes, the protein product of DQA2 and DQB2 have not been identified, while DQB3 is a pseudogene. The DR locus has two B genes and one A gene. The DRB2 is a psuedogene. Pseudogenes may serve as donors of nucleotide sequences that can be used in generating additional Class II polymorphism.

The genes for the transporter proteins are located between the DP and DQ gene loci. These proteins are involved in the antigen peptide transport across the endoplasmic reticulum.

There are multiple point mutations that occur on these loci due to the single nucleotide substitutions and reciprocal recombination. There is also the phenomenon of gene conversion that occur due to small blocks of genetic material that are exchanged between Class II genes in a non-reciprocal manner. MHC genes have the highest mutation rate of all germ line genes. The expression of Class I and Class II MHC molecules on the surface of various cells is controlled by various cytokines, such as interferon γ (INFγ) and Tumour Necrosis Factor (TNF).

6.1.4 Class III MHC Genes

Class III MHC genes encode proteins of complement system, soluble serum proteins and tumour necrosis factors. These proteins have a wide variety of functions that are not directly linked to antigen presentation.

Genes coding for Class III MHC molecules are located in the Class III region on chromosome 6, between the HLA-B locus and the HLA-DR sub region. These Class III molecules are not histocompatibility antigens, but are certain soluble molecules such as complement components and tissue necrosis factor.

Mouse H-2 Complex

Complex	H–2						
MHC Class	I	II		III		I	
Region	K	IA	IE	S		D	
Gene products	H–2K	IA αβ	IE αβ	C′ proteins	TNF-α TNF-β	H–2D	H–2L

Human HLA Complex

Complex	HLA							
MHC Class	II			III		I		
Region	DP	DQ	DR	C4, C2, BF		B	C	A
Gene products	DP αβ	DQ αβ	DR αβ	C′ proteins	TNF-α TNF-β	HLA-B	HLA-C	HLA-A

Figure 6.1 Simplified organization of the major histocompatibility complex (MHC) in the mouse and human.

6.1.5 Murine Class I Loci

The mouse MHC is called the H-2 locus. The regions encoding the Class I and Class II genes are given letter designations. MHC Class I molecules are encoded by H-2K region. The MHC gene and gene loci are highly polymorphic—they vary between different strains both in the numbers of genes and in their structures.

The mouse MHC (H-2) has three Class I loci, but the number of Class I gene varies between haplotypes.

6.2 THE MAJOR HISTOCOMPATIBILITY COMPLEX (MHC) ANTIGENS

The genetic loci involved in rejection of foreign or non-self tissues form the Major Histocompatibility Complex (MHC). The highly polymorphic cell surface structures involved in rejection are known as MHC antigens. The human MHC is known as the Human Leukocyte Antigens (HLA) system. These molecules are characterized as antigens that allow alloantibodies to bind and destroy leucocytes. Although they are known as antigens, they are only recognized as such when exposed to a non-self immune system, like following organ transplantation.

Three classes of molecules (I, II and III) have been identified as encoded both within the murine and human MHCs. Class I and Class II molecules present distinct structural entities; although multiple Class I and Class II genes exist within the MHC, all Class I and Class II genes products have similar overall structure. Class III region contains a diverse collection of over 20 genes, including some that encode complement system molecules (C4, C2 and factor B), and some that are involved in the processing of antigen. There are no functional or structural similarities between Class III gene products and the Class I or Class II molecules. Only Class I or Class II genes and their gene products are involved in triggering T-lymphocytes into activity.

The folding of an MHC molecule forms a **peptide-binding cleft**. This accommodates peptides that have been processed by the cell, to be presented to T-cells. Peptides of 8 or 9 residues can bind to Class I molecules, whereas longer peptides can bind to Class II molecules (Table 6.2).

Table 6.2 Peptide Binding by Class I and Class II MHC Molecules

	Class I Molecules	*Class II Molecules*
Peptide-binding domain	$\alpha 1/\alpha 2$	$\alpha 1/\beta 1$
Nature of peptide-binding cleft	Closed at both ends	Open at both ends
General size of bound peptides	8–10 amino acids	13–18 amino acids
Peptide motifs involved in binding to MHC molecule	Anchor residues at both ends of peptide; generally hydrophobic carboxyl-terminal anchor	Anchor residues distributed along the length of the peptide
Nature of bound peptide	Extended structure in which both ends interact with MHC cleft, but middle arches up away from MHC molecule	Extended structure that is held at a constant elevation above the floor of MHC cleft

Binding pockets within the clefts are able to accommodate different peptides depending on the haplotype. Since MHC molecules are highly polymorphic, and since a cell can express several different MHC molecules, the cell can present many different antigenic peptides to a T-cell.

6.2.1 The Structure of MHC Class I Molecules

Class I MHC molecules are gene products that are expressed on the cell surface of all nucleated cells within the body as well as on platelets, showing highest concentration on T-lymphocytes, B-lymphocytes, and macrophages.

The structure of MHC Class I molecules (Figure 6.2) comprises a glycosylated heavy-chain (45-kDa) non-covalently associated with β_2-microglobulin (12-kDa), a polypeptide that is also found free in serum.

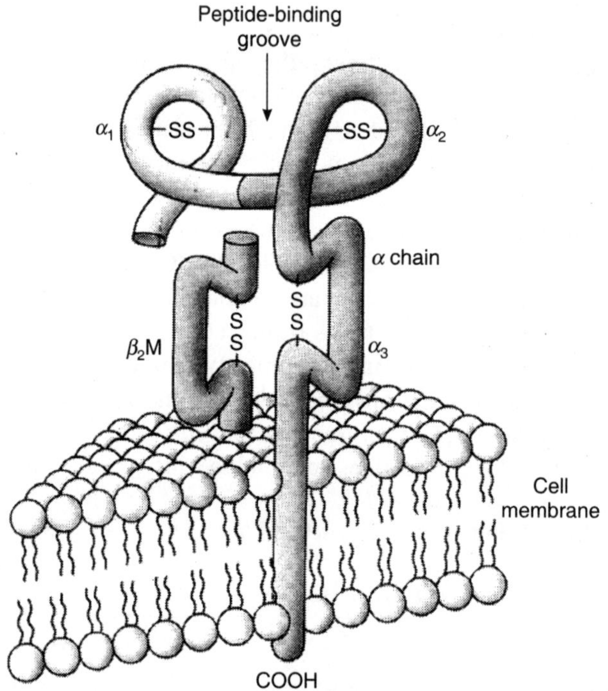

Figure 6.2 Simplified view of the structure of a MHC Class I antigen.

It consists of a large α chain noncovalently associated with a small β2-microglobulin molecule. The α genes within the A, B, and C regions of human HLA complex encode the chain and it is a polymorphic transmembrane glycoprotein of 45 kD. β2-microglobulin is coded by gene located on different chromosome and is an invariant protein of 12 kD. The expression of Class I MHC molecule on the cell surface needs the association of the α chain and β2-microglobulin. The α chain is anchored to the plasma membrane by inserting its hydrophobic transmembrane segment and hydrophilic cytoplasmic tail. The α chain is organized into three external domains of α1, α2 and α3, each of which contains approximately 90 amino acids. The transmembrane domain has 40 amino acids and the cytoplasmic anchor segment is made up of 30 amino acids. There is a considerable homology between α3, β2-microglobulin and the constant region domain of immunoglobulin. It is therefore called the immunoglobulin superfamily. In size and organization the β2-microglobulin is similar to α3 domain.

The peptides that binding to the peptide by the cleft of Class I MHC molecules are usually 8 to 10 amino acids residues long. The peptide-binding cleft is situated on the top of the MHC molecule and is made up of $\alpha 1$ and $\alpha 2$ domains. The $\alpha 3$ domain is highly conserved among Class I MHC molecules and contain a sequence that is recognized by the CD8 membrane receptor of the T_C-cells.

The specificity of the HLA molecule resides within its extracellular domains that serve as antigen binding sites for previously processed peptides by antigen-presenting cells to which HLA is bound. β_2-microglobulin (β_2-m) is essential for expression of MHC Class I molecules. It is non-polymorphic in man, but is dimorphic in mice. It has the structure of an immunoglobulin constant domain. This molecule also associates with a number of other Class I related structures. A long groove separates the α helices of the α_1 and α_2 domains. The majority of polymorphic residues and T-cell epitopes on Class I molecules are located in or near the groove.

MHC Class I molecules are membrane-bound and extend through the cell membrane into the cytoplasm. They are associated with three main functions, which are:

- serving as specific tissue markers
- recognition of foreign antigens
- graft rejection

6.2.2 The Structure of MHC Class II Molecules

Class II molecules (Figure 6.3) resemble Class I molecules in the overall structure. These are products of the Class II genes (A and E in the mouse; DR, DQ and DP in humans). The MHC Class II binding groove accommodates longer peptides.

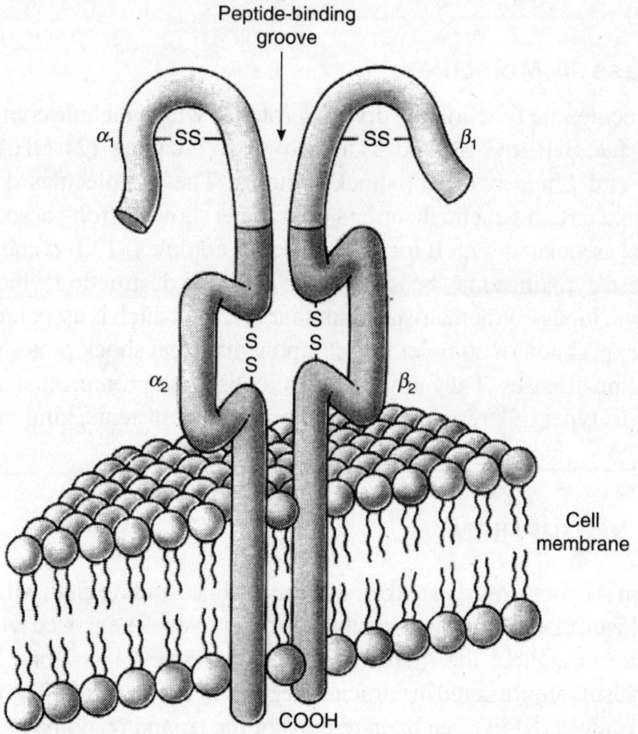

Figure 6.3 Simplified view of the structure of a MHC Class II molecule showing its heterodimeric structure.

Class II MHC molecule is a heterodimer consisting of two different polypeptide chains, an α chain of 33 kD and a β chain of 28 kD. These chains associate with each other noncovalently. Each chain contains two extracellular domains, one connecting peptide, one transmembrane segment and a cytoplasmic tail or anchor segment. The external domain consists $\alpha 1$ and $\alpha 2$ domains and $\beta 1$ and $\beta 2$ domains, and bear sequence homology to the immunoglobulin fold domain structure. Hence Class II molecules are classified as immunoglobulin superfamily. The top of a Class II molecule is composed of the $\alpha 1$ and $\beta 1$ domains and forms the antigen binding cleft for processed antigen. A third chain called γ or invariant chain (CD74) of molecular weight 32 kD is associated with intracellular Class II molecules. This is not coded for by genes located within the MHC.

The extracellular region of the chain is the binding site for attachment of processed antigenic peptides fragments. Its expression is modulated by IFN-γ and IL-4.

Class II MHC molecules have great polymorphism. The amino acids show great variability, that is restricted to the $\alpha 1$ and $\beta 1$ domains where there are 3 or 4 discrete hypervariable regions. $\alpha 1$ and $\beta 1$ are the outermost domains that fold together to form an open ended groove, which acts as the antigen-binding site. The hypervariable regions are located in the walls of the groove.

Class II MHC molecules are located on a surface of immunocompetent cells, including monocytes, B-lymphocytes, activated T-lymphocytes, and macrophages. The number of Class II molecules on the cell surface is enhanced in rapidly dividing cells and in cells treated with interferon. Although cells like resting T-cells, endothelial cells, and thyroid cells do not normally express Class II molecules, they may be induced to do so.

The main function of Class II MHC molecules is their participation in the recognition of foreign antigens.

6.2.3 MHC Class III Molecules

MHC Class III molecules are functionally diverse proteins, which include complement components C2, C4a, C4b and factor B, two steroid 21 hydroxylase enzymes (21-OHA, 21-OHB), tumour necrosis factors α and β and two heat shock proteins. These molecules do not have a role in antigen presentation. Certain genetic disorders of humans show a strong association with Class III locus. Spondylitis is associated with B locus of the HLA complex. TNF α and β genes are found at this B locus and these cytokines may be involved in cartilage destruction which is characteristic of this disease. Systemic lupus erythematosus is another disease which is associated with MHC alleles through defects in expression of complement components. Heat shock proteins may also be linked to certain autoimmune diseases. They are a group of conserved proteins that are produced by cells in response to various types of stress like lack of oxygen and nutrients, viral infection of a cell, and heat.

6.3 MHC POLYMORPHISM

MHC polymorphism is concentrated in and around the peptide-binding cleft. MHC has got an extreme degree of polymorphism, that is, structural variability of the molecules encoded within it. Polymorphism is not evenly spread throughout the MHC. Within a particular Class I or Class II molecule, the structural polymorphisms are clustered in particular regions of the molecule. The amino acid variability in Class I antigens is clustered in three main regions of the α_1 and α_2 domains.

Most of the polymorphic amino acids in Class I and Class II antigens are clustered on the top of the molecule in the large groove that acts as the peptide-binding site. Thus, variation is almost exclusively centred in the base of the antigen-binding groove or pointing in from the sides of the α-helix region.

The enormous diversity in MHC molecules results from polymorphism due to the presence of multiple alleles at the given genetic locus within a given species. Different MHC molecules are expressed by different individuals of the same species. The number of amino acid differences between MHC alleles can be up to 20 amino acid residues contributing to the uniqueness of each allele. Human Class I molecule has 59 A alleles, 111 B alleles and 37 C alleles. This creates a major hurdle when it comes to matching MHC molecules for successful organ transplants. Several allelic MHC molecules encoded by a single locus show a sequence divergence of 5 to 10 per cent. The sequence variation among MHC molecules is not randomly distributed along the entire length of polypeptide. Instead it is clustered in short stretches mainly within the $\alpha1$ and $\alpha2$ domains of Class I molecule and $\alpha1$ and $\beta1$ domains of Class II molecules.

6.4 RECOGNITION OF ALLOANTIGENS

In humans several different allelic variants of Class I and II MHC molecules have been identified. At any given time, an individual is supposed to express up to six different Class I molecules and up to 12 different Class II molecules. Yet this limited number of molecules is capable of presenting a huge variety of antigens to the T-cells and induces cellular and humoral immune responses against them. MHC molecules do not show specificity and therefore, can bind to numerous different peptides, and some peptides can bind to several different MHC molecules.

Recognition of donor alloantigens that are genetically dissimilar antigens, by the immune system and subsequent rejection of cells bearing these incompatible antigens is accomplished through the recipient's T-Cell Receptors (TCRs) and occurs in two ways. This is done by:

- Recognition of donor major histocompatibility antigens.
- Recognition of donor MHC bound peptides.

In allografts that differ from the host at both Class I MHC and Class II MHC, two sub-populations of host T-lymphocytes are activated during recognition of alloantigens in the following two ways:

- The host $CD4^+$ T-lymphocytes are activated by allogeneic Class II molecules located on the surface of Antigen-Presenting Cells (donor APCs). Once activated, the $CD4^+$ cells produce cytokines (co-stimulators), which activate other cells like cytotoxic $CD8^+$ T-lymphocytes, B-lymphocytes and macrophages. The newly cytokine-activated and proliferating cells can reject the MHC incompatible graft by specific (cell-mediated and humoral) and nonspecific (inflammatory reaction) immune responses.
- Alternatively, $CD8^+$ T-cells may be directly stimulated by allogeneic Class I MHC molecules expressed on the surface of all cells of a graft, with required helper T-cell participation.

6.4.1 The Peptides Bound to Class I MHC Molecules

The peptides that bind to Class I MHC molecules are nonamers with specific amino acid sequences that are essential for binding of the peptide to a particular MHC molecule. The nonameric nature of the peptide suits the size of the peptide binding cleft of Class I MHC. The peptides that bind to these molecules share similar amino acids residues at several defined positions along the peptide which are responsible for fitting the peptide in the groove of the MHC molecule. These conserved amino acid residues are called anchor residues, which are generally hydrophobic residues and contain carboxyl terminal anchor. There are a few exceptions where charged amino acids are also seen. The majority of contacts between Class I MHC molecules and peptides involve residue 2 at the amino terminal end and residue 9 at the carboxyl terminus of the nonameric peptide. The Class I MHC bound peptide arches away from the middle of the cleft, but interact with the MHC at both ends. The peptides that are slightly longer or that have different middle residues, but the correct anchor residues at each end can be bound to the same MHC molecule. Arching away of the peptide bound to Class I MHC is advantageous because the arched regions are better exposed and hence, can directly interact with the T-cell receptors.

6.4.2 The Peptides Bound to Class II MHC Molecules

The peptides bound to class II MHC molecules are presented to CD4 T-cells. Class II MHC molecules can also bind to a variety of peptides. These peptides are generally derived from the exogenous proteins of pathogens that are degraded within the endocytic pathway. The peptides that bind to Class II MHC contain 13 to 18 amino acid residues. The peptide-binding cleft of Class II MHC is open on both ends and allow peptides to extend beyond the ends. The bound peptides have a roughly constant elevation on the floor of the binding cleft and also have conserved motifs. Class II binding peptides have hydrogen bonds distributed throughout the binding site. The peptides that bind to these molecules contain a core sequence of seven to ten amino acids, mainly aromatic or hydrophobic residues at the amino terminal and three additional hydrophobic anchors in the middle portion and carboxyl terminal end of the peptide.

Class I and II molecules act as receptors that bind antigen fragments and present the antigen to lymphocytes. They also regulate immune responses. The Class III genes code for a mixture of proteins with many functions that are not directly linked to antigen presentation.

The level of Class I MHC expression varies among different nucleated cell types with the highest level of expression seen in lymphocytes which is about 5×10^5 molecules per cell. On the other hand liver cells, fibroblasts, muscle cells and neural cells express very low levels of Class I MHC. Any particular MHC molecule can bind many different peptides. If a cell is normal, its Class I molecules will display self peptides of cell origin like histones, ribonucleoproteins and cytochrome c. On the other hand if a cell is transformed or infected with a virus, Class I MHC molecule will display viral peptides or altered self-peptides.

Class II MHC molecules are expressed only by the antigen presenting cells that includes macrophages, dendritic cells, B-cells and thymic epithelial cells. Expression of Class II MHC molecules by these cells, differ at different stages of development. Pre-B-cells do not express Class II molecules, but mature B-cells constitutively express them. Naive macrophages and monocytes express only low levels of these molecules, but on encountering an antigen their expression increases significantly. Since human Class II MHC gene contains three loci of DP, DQ and DR, a heterozygous individual expresses six parental Class II molecules and six molecules containing α and β chain combination from either parent. The number of Class II MHC molecules on the cell surfaces increases in rapidly dividing cells and in cells treated with IFN.

6.4.3 In Vitro Recognition and Response to Alloantigens (Mixed Leukocyte Reaction)

The most powerful immune response to genetically incompatible allografts is seen when the hosts $CD8^+$ and $CD4^+$ T-lymphocytes recognize dissimilar donor MHC Class I and MHC Class II antigens, respectively. The response of the immune system to incompatible tissue HLA antigens can be evaluated by Mixed Leukocyte Reaction (MLR). When differences in MHC antigens exist in two cultured cell populations, many mononuclear cells of one cell population are stimulated and proliferate on recognition of alloantigens present on the other cell population. The greater the incompatibility in the MHC antigens between the donor and recipient (quantitated by MLR), the greater is the potential for a graft rejection.

MULTIPLE-CHOICE QUESTIONS

1. Class I MHC molecules are found on:
(a) B-cells and macrophages
(b) T-cells only
(c) neutrophils, T-cells and B-cells
(d) all nucleated cells

2. Class II MHC molecules contain:
(a) one γ and one δ chain
(b) one α and one β chain
(c) two light and two heavy chains
(d) one γ chain

3. An individual's HLA haplotype is:
(a) a chromosome segment
(b) the individuals complete set of MHC alleles
(c) one allele at each MHC Class I locus
(d) one allele at each MHC Class II locus

4. β-2 microglobulin is an integral part of:
(a) IgM
(b) MHC Class I molecules
(c) MHC Class II molecules
(d) T-cell antigen receptor

REVIEW QUESTIONS

6.1 Why do animals need histocompatibility molecules?
6.2 Describe the structural function of Class II MHC molecules.
6.3 Describe the structural function of Class I MHC molecules.
6.4 What do you know about the organization of Class I and Class II MHC genes?
6.5 Discuss the genetic polymorphism of MHC molecules.

7

COMPLEMENT

7.1 THE COMPLEMENT SYSTEM

The complement system is a set of proteins that act together to destroy invading microorganisms. This system consists of a group of plasma and membrane proteins, which are normally present in the inactive form. When activated through a series of steps of either the classical pathway, the alternative pathway or the lectin pathway, these proteins are converted into active enzymes by proteolysis. The activation of these proteins results in the disruption of cell membranes and the destruction of cells or invading microorganisms. Jules Bordet discovered the complement system in 1893 as "alexine". Paul Ehrlich later termed it "complement".

The complement system lacks specificity. It has the capability of participating in both the natural and specific immune responses. The system is part of the innate immune system and has evolved mechanisms of self/non-self-discrimination. There are regulatory molecules present on host tissues that inhibit the activation of complement.

The term 'complement' was originally applied to describe the activity in serum, which could complement the ability of specific antibody to cause lysis of bacteria. This is a heat labile activity in serum.

This system is the primary humoral mediator of antigen-antibody reactions. The system consists of at least 20 plasma proteins capable of interacting with each other, with antibody and with cell membrane. They are all serum proteins and they make up about 10 per cent of the globulin fraction of serum. Their molecular weights vary between 244-kDa and 460-kDa. Serum concentrations in humans vary between 20 μg/ml of C2 and 1300 μg/ml of C3 (Table7.1).

Complement components are synthesized at various sites throughout the body. The components are designated by numerals C1, C2, C3, etc. and by names such as factor B or factor D. Most C3, C6, C8, and B are made in the liver; and C2, C3, C4, C5, B, D, P and I are synthesized by macrophages. As a result, these components are readily available for defense at sites of inflammation where macrophages accumulate. Genes located in the MHC Class III region code for B, C4, and C2.

Table 7.1 Complement Components

Name	*MW* (kDa)	*Serum Concentration* (μg/ml)
Classical pathway		
C1q	460	80
C1r	83	50
C1s	83	50
C4	200	600
C2	102	20
C3	185	1300
Alternative pathway		
D	24	1
B	90	210
Terminal components		
C5	204	70
C6	120	65
C7	120	55
C8	160	55
C9	70	60
Control proteins		
C1-INH	105	200
C4-bp	550	250
H	150	480
I	88	35
P	4 × 56	20
Vitronectin	83	500

There are three separate enzyme pathways within the complement system: the classical pathway initiated by antibody bound to an antigen; the alternative and lectin pathways initiated by foreign surfaces; and the terminal pathway that results in the destruction of microorganisms. Once activated, the active enzyme is capable of cleaving and activating the next protein in the activation pathway, thus creating a cascade effect. These proteins then produce several effects associated with humoral immune responses. These include immune adherence, opsonization, chemotaxis, kinin activation, cell lysis, and inflammation.

7.2 ACTIVATION

There are three interrelated pathways of enzyme cascades involved in the activation of complement. These are the classical pathway, the alternative pathway and the lectin pathway. Different substances trigger each pathway. The three pathways converge at a point from which a common final pathway begins to terminate at the end point of complement activation, which is cytolysis or cytotoxicity. The activation of any of these pathways produces biologically active fractions of complement, which serve as effector molecules in the various defense mechanisms of the natural and specific immune system.

The two main pathways for complement activation, the classical and alternative pathways, reflect the innate and adaptive immune response. Both pathways lead to the formation of a convertase that cleaves C3 to C3a and C3b. The classical pathway links the adaptive immune system through the binding of immune complexes to C1q. The alternative pathway (innate) is activated by the chance binding of C3b to the surface of a microorganism. Complement has a role both in defense against bacteria, and also in the disposal of immune complexes, which would otherwise lead to autoimmunity and immune-complex diseases.

7.2.1 Activation of Complement by the Classical Pathway

Activation of complement starts with the C1 component (Figure 7.1).

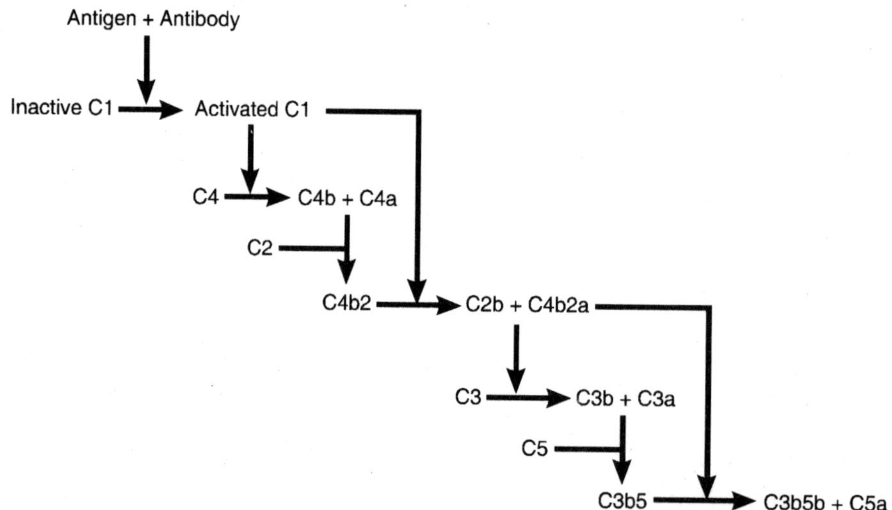

Figure 7.1 The classical complement pathway to form C3 convertase.

C1 is a pentamolecular, Ca^{++} dependent, protein complex consisting of a single C1q molecule, two C1r and two C1s molecules. C1q can bind an immunoglobulin attached to an antigen, but not to free immunoglobulin molecules. The site on the IgG molecule that binds C1q is located on its C_H2 domain, while in IgM it is on C_H3 domain. It is only when an immunoglobulin binds the antigen that the complement-activating site is exposed to C1q, which causes one of the C1r molecules to activate itself (auto catalysis) and then the other C1r, to yield two active C1r enzymes. These two enzymes then cleave the two C1s molecules to give active $\overline{C1s}$ serine esterases. The stimuli that can activate C1 include single antigen-bound molecules of IgM or paired antigen-bound molecules of IgG. Multiple binding of the globular domains of C1q to complexed IgG or IgM leads to a conformational change in the C1 complex. Activated C1 acts on C4 and C2, which in turn act on C3. Activated $\overline{C1s}$ acts on C4 cleaving off a small portion from the N-terminus, called

C4a and leaving the major fragment C4b that binds to any nearby cell membrane. Surface bound C4b acts as the binding site for the zymogen C2. The bound C2 is a substrate for C$\overline{1s}$ and is cleaved to release C2b, while the large C2a remains bound to C4b to form C$\overline{4b2a}$, the classical pathway C3 convertase enzyme. The C2 must be bound to the C4 before it can be cleaved. This substrate modulation confines the activities of proteases such as C1s to the targets under attack. The C2a part of the C$\overline{4b2a}$ complex is a protease, and it can activate C3. The C$\overline{4b2a}$ complex is therefore called classical **C3 convertase**.

The classical C3 convertase acts on C3 the most important of the complement components, to cut off a small fragment called C3a. This is a major step, since each C$\overline{4b2a}$ complex can activate up to two hundred C3 molecules.

7.2.2 The Alternative Pathway of Complement Activates Spontaneously

A second route called the alternative complement pathway can also activate C3 (Figure 7.2). The alternative pathway, earlier known as the **properdin** pathway, bypasses the need for C1, C4 and C2. For activating this pathway, the presence of C3b is required, which is continuously generated. Several factors involved in the system have been isolated and identified. They include properdin, and factors A, B, D, H and I.

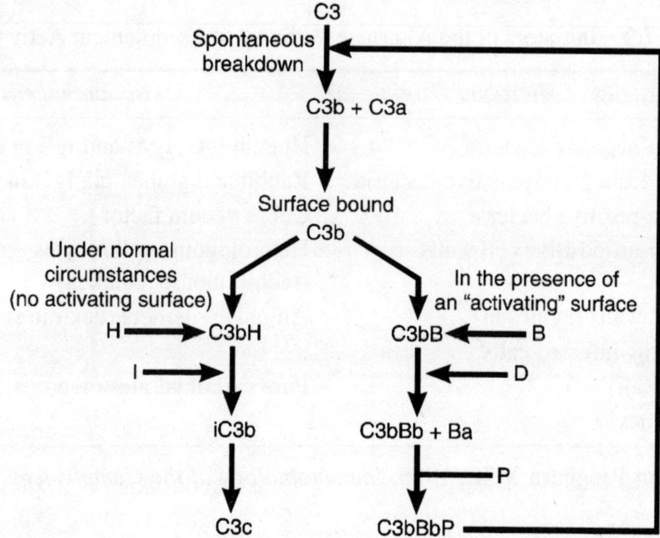

Figure 7.2 The alternate pathway of complement activation.

Native C3 can break down slowly, but spontaneously in the plasma, so that small amounts of C3b are continuously generated in normal serum. C3b can bind to nearby cell surfaces. The cell-bound C3b either may be inactivated or may initiate the alternative pathway of complement activation, depending upon the affinity of C3b for a protein called factor H.

When factor H interacts with sialic acid and other neutral or anionic polysaccharides on cell surfaces, its binding to surface-bound C3b is enhanced up to 100 fold. Activating surfaces that inhibit factor H and therefore, permit activation of C3b include cell walls from many bacteria, polyanions such as bacterial lipopolysaccharides and nucleic acids. Small polysaccharides, viruses, aggregated immunoglobulins and foreign particles like asbestos also inhibit factor H. Antibodies may permit the activation of the alternative pathway on cell membranes by masking sialic acid residues. In presence of activating surfaces, the binding of factor H is reduced and the C3b is free to bind factor B.

In the absence of an activating surface, factor H binds to C3b resulting in alteration of C3b so that it becomes susceptible to an enzyme called factor I. This results in accelerating the decay of C3b by factor H. In a normal, stable situation, factors H and I destroy C3b as soon as it is generated, and the alternative pathway for C3-convertase activation remains inactive.

In the presence of an activating surface, the ability of factor H to bind C3b is greatly reduced and the C3b can bind to factor B. This complex C3bB has weak C3 convertase activity. The bound factor B is cleaved by factor D into a soluble fragment Ba, and a bound fragment called Bb. The complex $\overline{C3bBb}$ has potent C3 convertase activity. Factor B is synthesized in liver and in macrophages. Factor D, which is synthesized by monocytes, cleaves B after it is bound to C3b. The alternate C3 convertase, $\overline{C3bBb}$ has a half-life of only five minutes. If factor P (also called properdin) binds the complex, the half-life of $\overline{C3bBbP}$ is increased to 30 minutes.

The alternative pathway is an independent means to generate C5b and is therefore a component of the innate immune system. There are many initiators of the complement activation through this pathway (Table 7.2).

Table 7.2 Initiators of the Alternate Pathway of Complement Activation

Pathogens and Particles of Microbial Origin	*Nonpathogens*
Many strains of gram-negative bacteria	Human IgG, IgA, and IgE in complexes
Lipopolysaccharides from gram-negative bacteria	Rabbit and guinea pig IgG in complexes
Many strains of gram-positive bacteria	Cobra venom factor
Teichoic acid from gram-positive cell walls	Heterologous erythrocytes (rabbit, mouse, chicken)
Fungal and yeast cell walls (zymosan)	Anionic polymers (dextran sulphate)
Some viruses and virus-infected cells	
Some tumour cells (Raji)	Pure carbohydrates (agarose, insulin)
Parasites (trypanosomes)	

Source: Adapted from Pangburn M.K., 1986, *Immunobiology of the Complement System*, Academic Press, Orlando.

Thus, the classical pathway convertase is the combination of C4 with C2 forming C4b2a, and the alternative pathway convertase is the combination of C3 with factor B forming C3bBb. The C3b generated by these two enzymes binds to the target membrane and becomes a focus for further C3b production, thus making it an amplification loop.

Proteolytic cleavage of C3a from C3 by C3 convertase results in the binding of C3b to proteins and sugars in the immediate vicinity of the activation site. The bound C3b acts as a focus for further complement activation through the alternative pathway amplification loop.

Both kinds of C3 convertase can be turned into a C5 convertase which catalyses the first step in a cascade that leads to the production of membrane attack complex.

7.2.3 The Lectin Pathway

Complement cascade can be activated by the lectin pathway too. Lectins are proteins that bind to a carbohydrate. This pathway does not depend on antibody for its activation. The pathway is activated by the binding of Mannose-Binding Lectin (MBL) to mannose residues on glycoproteins or carbohydrates on the surface of microorganisms. MBL is an acute phase protein produced in inflammatory responses. After MBL binds to the surface of a cell or pathogen, an MBL-Associated Serine Protease (MASP) binds to it. The active complex thus formed causes cleavage and activation of C4. MASP is similar to C1r and C1s in structure and function. After initiation, this pathway proceeds through the action of C4 and C2, to produce a C5 convertase. The ways by which C3 and C5 convertases are formed are given in Table 7.3.

Table 7.3 Complement Components in the Formation of C3 and C5 Convertases

	Classical Pathway	*Lectin Pathway*	*Alternative Pathway*
Precursor proteins	C4 + C2	C4 + C2	C3 + factor B
Activating protease	$\overline{\text{C1s}}$	MASP	Factor D
C3 convertase	$\overline{\text{C4b2a}}$	$\overline{\text{C4b2a}}$	$\overline{\text{C3bBb}}$
C5 convertase	$\overline{\text{C4b2a3b}}$	$\overline{\text{C4b2a3b}}$	$\overline{\text{C3bBb3b}}$
C5-binding component	C3b	C3b	C3b

7.3 TERMINAL COMPLEMENT PATHWAY

Formation of the Membrane Attack Complex (MAC) constitutes the final phase of complement activation. This is done by the enzymatic cleavage of C5. C5 must be bound to C3b before the C5 convertase enzyme can cleave it. The classical pathway C5 convertase enzyme is a trimolecular complex, $\overline{\text{C4b2a3b}}$ in which C3b is covalently bound to C4b. The alternative pathway C5 convertase enzyme is also a trimolecular complex, $\overline{\text{C3bBb3b}}$, in which one C3b is covalently bound to the other. Cleavage of C5 releases the small peptide fragment C5a, which is a potent anaphylatoxin.

The non-enzymatic assembly of C5b-9 forms the membrane attack complex. C5b binds C6, forming C5b6 and this then binds C7 to form a C5b67 complex. The binding of C7 to C5b6 complex can occur either in solution or bound to the cell membranes. When C5b67 is formed, it generates a membrane-binding site. If it does not bind to a membrane, the complex self-aggregates and is inactivated. This complex is largely hydrophobic that preferentially inserts into lipid bilayers. C8, which is a three-chain molecule binds to this active cell-membrane bound C5b67 complex, and then undergoes a conformational change and penetrates the lipid bilayer of the cell.

The entire C5b678 complex induces polymerization of C9. Up to eighteen C9 molecules come together to form a tubular structure called the Membrane Attack Complex (MAC). The MAC can insert itself into the cell membrane forming a transmembrane channel in the membrane and osmotic hydrolysis of the cell occurs (Figure 7.3). Although a small amount of lysis occurs when C8 binds to C5b67, it is the polymerized C9 that causes the majority of lysis.

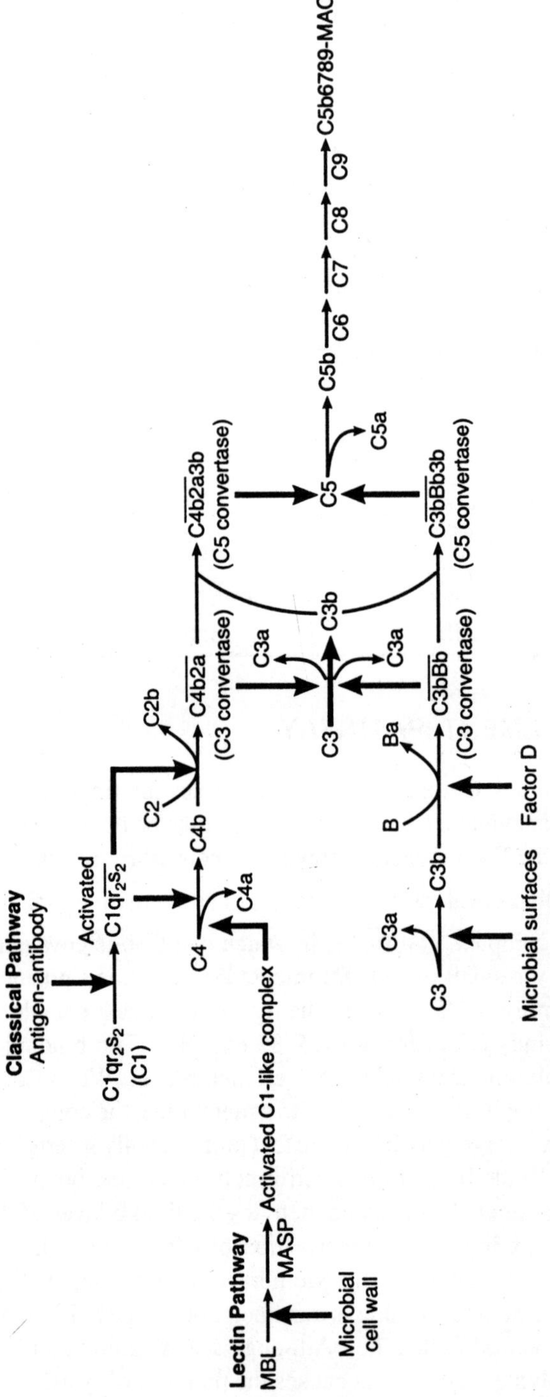

Figure 7.3 Overview of the complement activation pathways.

7.4 EFFECTS

Activation of the complement system results in inflammation, chemotaxis, opsonization, and cell lysis. It also regulates the immune system. The main consequences of complement activation are **biological effects** (opsonization, chemotaxis) and **cell lysis.**

7.4.1 Biological Effects of Complement

Biological effects of complement fragments (Figure 7.4) are through their binding to specific complement receptors on various cell surfaces. It includes bacterial opsonization and chemotaxis.

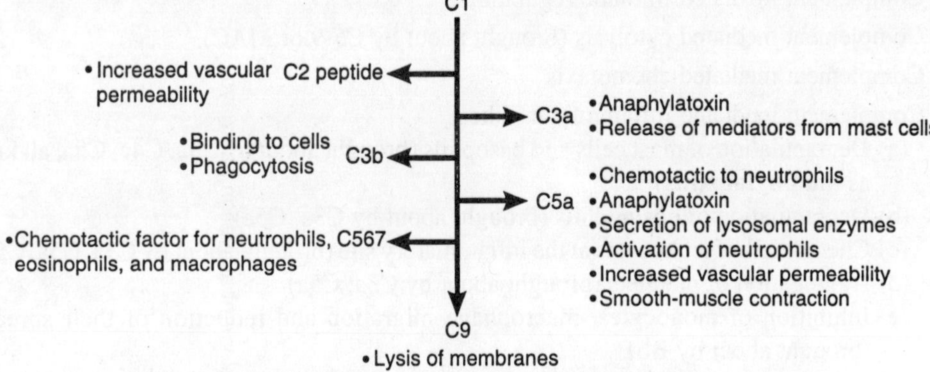

Figure 7.4 The biological consequences of complement activation.

The major beneficial activities of complement are:

(i) Promotion of killing of microorganisms
(ii) Clearing of immune complexes
(iii) Induction and enhancement of antibody responses

Complement may cause harm to the host under certain circumstances if activated systemically on a large scale. This activation may also occur by tissue necrosis, e.g. during myocardial infarction or by autoimmune response to host tissues.

7.4.2 Cell Lysis

Cell lysis is an important defense mechanism developed by the humoral immune system against microbes. Specific antibodies against microbial antigen are produced which bind to the antigen forming an antigen-antibody complex on its surface. The antigen-antibody complex activates the classical complement pathway and produces a variety of biologically active complement fragments. These fragments produce cell lysis as well as other biological effects. The final sequence of the activated classical pathway produces complement components (C5 through C9) known as the Membrane Attack Complex (MAC). MAC inserts into the surface membrane of the target cell or organism that causes the antigen-antibody complex, causing cell lysis and its destruction. When MAC inserts into the surface membrane of the target cell, pores are formed that allow passage of water and small molecules (e.g. Ca^{++} ions) out of the cytoplasm. The large molecules remain within, creating a concentration gradient that results in an entry of water molecules into the cytoplasm to maintain equilibrium between the interior and exterior of the cell. The entry of water molecules into the cytoplasm causes swelling and cell lysis.

7.5 CONSEQUENCES OF COMPLEMENT ACTIVATION

The **consequences of complement activation** may be summarized as under:

1. Complement mediated opsonization of particulate antigens, increasing their phagocytosis (brought about by C3b, C4b)
2. Solubilization and removal of immune complexes (brought about by C3b)
3. Complement mediated immune regulation
4. Complement mediated cytolysis (brought about by C5-9 or MAC)
5. Complement mediated chemotaxis
6. Complement mediated inflammation, like:
   - (a) Degranulation of mast cells and basophils (brought about by C3a, C4a, C5a, all known as **anaphylatoxins**)
   - (b) Degranulation of eosinophils (brought about by C3a, C5a)
   - (c) Chemotaxis of leukocytes at the inflammatory site (brought about by C3a, C5a, C5b67)
   - (d) Aggregation of platelets (brought about by C3a, C5a)
   - (e) Inhibition of monocytes, macrophage migration and induction of their spreading (brought about by Bb)
   - (f) Release of neutrophils from bone marrow (brought about by C3c)
   - (g) Release of hydrolytic enzymes from neutrophils (brought about by C5a)
   - (h) Increased expression of complement receptors types 1 and 3 (brought about by C5a)
7. Viral neutralization (brought about by C3b, C5b-9)

7.6 COMPLEMENT FIXATION TEST

This is a serologic test that uses complement to measure antibody levels in serum. This test is performed in two parts. First, antigen and the serum under test (deprived of its complement by heating at 56°C) are incubated in the presence of normal guinea pig serum, which provides a source of complement. After the antigen-antibody-complement mixture is allowed to react for a short period, the amount of free complement remaining is measured by adding an *indicator system* consisting of antibody-coated sheep RBC. Lysis of these red blood cells is a negative result, as it indicates that complement was not *fixed* and antibody was therefore absent from the serum under test. On the other hand, absence of lysis is a positive result. The titer may be considered to be the reciprocal of the highest dilution of serum by which 50 per cent of the red blood cells are lysed. In this test, using the correct amount of complement is critical as too little complement results in incomplete lysis, whereas excessive complement is not completely activated by the immune complexes and therefore, leads to false negative results. Thus, excessive antigen interferes with complement activation, while insufficient antigen may fail to fix the complement well.

7.7 REGULATION

The complement system is carefully regulated as its activation is very significant and, therefore, should not act in an uncontrolled fashion (Table 7.4).

Table 7.4 Proteins that Regulate Complement System

Protein	*Type of Protein*	*Pathway Affected*	*Immunologic Function*
C1 inhibitor (C1Inh)	Soluble	Classical	Serine protease inhibitor: causes $C1r_2s_2$ to dissociate from C1q
C4b-binding protein (C4bBP)*	Soluble	Classical and lectin	Blocks formation of C3 convertase by binding C4b; cofactor for cleavage of C4b by factor I
Factor H*	Soluble	Alternative	Blocks formation of C3 convertase by binding C3b; cofactor for cleavage of C3b by factor I
Complement-receptor type I (CR1)* Membrane-Cofactor Protein (MCP)*	Membrane bound	Classical, alternative, and lectin	Block formation of C3 convertase by binding C4b or C3b; cofactor for factor I-catalyzed cleavage of C4b or C3b
Decay-Accelerating Factor (DAF)*	Membrane bound	Classical, alternative, and lectin	Accelerates dissociation of $\overline{C4b2a}$ and $\overline{C3bBb}$ (classical and alternative C3 convertases)
Factor I	Soluble	Classical, alternative, and lectin	Serine protease: cleaves C4b or C3b using C4bBP, CR1, factor H, DAF, or MCP as cofactor
S protein	Soluble	Terminal	Binds soluble C5b67 and prevents its insertion into cell membrane
Homologous Restriction Factor (HRF) Membrane Inhibitor of Reactive Lysis (MIRL)	Membrane bound	Terminal	Bind to C5b678 on autologous cells, blocking binding of C9

*Proteins that are regulators of complement activation. All such proteins are coded on chromosome 1 in humans.

Regulation is done at the following levels:

1. Cell bound IgG, IgM and Immune complexes. These bind C1q and activate C1r and C1s, resulting in initiation of recognition phase.
2. C1 inhibitor of the classical pathway. This is a 105 kDa protein that controls the assembly of C4b2a (C3 convertase) by binding and blocking the activities of active C1r and C1s.
3. C1s. Acts enzymatically on C4 and C2 to form C3 convertase; initiates enzymatic phase.

4. C3 convertase. Shows enzymatic action on C3 to form C5 convertase. Mediates the progression of enzymatic phase.
5. C5 convertase that acts enzymatically on C5 and sets the stage for attack phase.
6. Factor H, which regulates the activities of C3b and C4b.
7. C4 binding protein is a heat labile protein that binds C4b and inhibits the formation of the $\overline{\text{C4b2a}}$ convertase. It also accelerates the decay and dissociation of C2a from C4b and destroys its activity.
8. Decay-accelerating factor (CD55) is a glycoprotein attached to red blood cells, neutrophils, lymphocytes, monocytes, platelets, and endothelial cells. It can bind to the classical ($\overline{\text{C4b2a}}$) or the alternate ($\overline{\text{C3bBb}}$) convertases and accelerate their decay when these molecules bind to a cell surface.
9. C5b attaches to membrane and initiates the formation of MAC; it therefore initiates the attack phase.
10. S protein is a plasma protein which binds soluble C5-7 complex and prevents the lysis of cells of surrounding tissues.
11. Protectin (CD59). It acts in the final stages of MAC formation by competing with C9 for a binding site on C8. It blocks the insertion of C5b67 into the cell membrane and so inhibits C9 polymerization.
12. Membrane cofactor protein (CD46) controls C3 turnover on cells by acting as a cofactor for the cleavage of C3b or C4b by factor I.
13. Finally, target cells like neutrophils can resist lysis by removing the MAC by endocytosis or exocytosis. These cells form membrane vesicles, that contain MAC, which are either shed into extracellular fluid or ingested and destroyed.

MULTIPLE-CHOICE QUESTIONS

1. Anaphylatoxins are produced by:
(a) C1 (b) C3
(c) C6 (d) Properdin

2. For serological tests the complement is derived from:
(a) horses (b) guinea pigs
(c) cattle (d) rabbits

3. Chemotactic properties are shown by the complement component:
(a) C1a (b) C2a
(c) C4a (d) C5a

4. For activating the classical complement pathway there is a requirement of interaction of antigen with:
(a) at least two IgG molecules (b) one IgA molecule
(c) at least two IgA molecules (d) one IgG molecule

5. In serum the complement component with the highest concentration is:
(a) C1 (b) C2
(c) C3 (d) C4

6. The alternative complement pathway is inhibited by:
(a) factor H (b) factor B
(c) factor P (d) factor D

7. C9 is a:
(a) cytokine (b) perforin
(c) acute phase protein (d) addressin

REVIEW QUESTIONS

7.1 Give an outline of the alternative complement pathway.

7.2 Give an account of the classical complement pathway.

7.3 Give a brief account of lectin to pathway of complement.

7.4 Write notes on:
(i) Membrane attack complex
(ii) Consequences of complement activation
(iii) MASP

8

Tolerance and Immunosuppression

8.1 TOLERANCE

Tolerance to self-antigens is an essential prerequisite for the functioning of immune system. It is the process by which the body makes sure that the immune responses are directed against the foreign antigens, and that the immune system does not react to normal self-antigens. The role of the immune system is to attack and destroy invading organisms and yet not attack the body. There is an absolute need to prevent the immune system from attacking the normal body tissues. Immunological tolerance is a state of unresponsiveness, which is specific for a particular antigen that is induced by a prior exposure to the antigen. The most important aspect of tolerance is self-tolerance, which prevents the body from mounting an immune attack against its own tissues. There is always a potential for such an attack because the immune system randomly generates a wide variety of antigen specific receptors, some of which may be self-reactive.

Discrimination between self and non-self is learnt during development. Immunological self must encompass all epitopes encoded by the individual's DNA, and the other epitopes are all considered as non-self.

It is not the structure of a molecule per se that determines whether it will be distinguished as self or non-self. Factors other than the structural characteristics of an epitope are also important. Among these are:

1. The stage of differentiation when lymphocytes first confront their epitopes.
2. The site of the encounter.
3. The nature of the cells presenting epitopes.

4. The number of lymphocytes responding to the epitopes.
5. The production of 'co-stimulatory' molecules by these cells.

Processes that occur during development when self and non-self discrimination is learnt prevent self-reactivity. There are several mechanisms that ensure that self-tolerance is established. Since both T- and B-cells generate antigen-binding receptors at random, the production of self-reactive cells cannot be prevented. Tolerance induction must therefore require selective elimination of those lymphocytes that develop receptors that can bind self-antigens.

Tolerance can be induced artificially. Antibodies to co-receptor and co-stimulatory molecules induce tolerance to transplants. Soluble antigens readily induce tolerance. Oral administration of antigens induces tolerance by a variety of mechanisms. Extensive clonal proliferation can lead to exhaustion and tolerance. Anti-idiotypic responses can be associated with tolerance. Tolerance in vivo relates to persistence of antigen.

Both B- and T-cells can be made tolerant, although T-cells are made tolerant considerably easier than B-cells. Involving both, the selection of T-cells in the thymus whose receptors do not bind self-antigens, and stimulation of T-cells that can react with foreign antigens induces the T-cells tolerance. One common way by which T- and B-cells can be made tolerant is through apoptosis induced by an incomplete stimulating signal.

Tolerance is the specific immunological unresponsiveness of an individual to an antigen. This tolerance is directed against antigens from normal tissues. However it can be established against any antigen by the use of appropriate procedures, that include administration of antigen to very immature animals, administration of very large or very small doses of antigen, or administration of antigen in a way that it cannot be processed properly.

The development of tolerance is associated with the stage of development of antigen sensitive cells when they first encounter antigen, and lymphocytes are trained not to respond to an antigen as a result of interaction with that antigen during foetal life.

The susceptibility of T- and B-cells to tolerance induction differs considerably. T-cells develop tolerance rapidly and easily, and they remain in that state for a longer time (Figure 8.1).

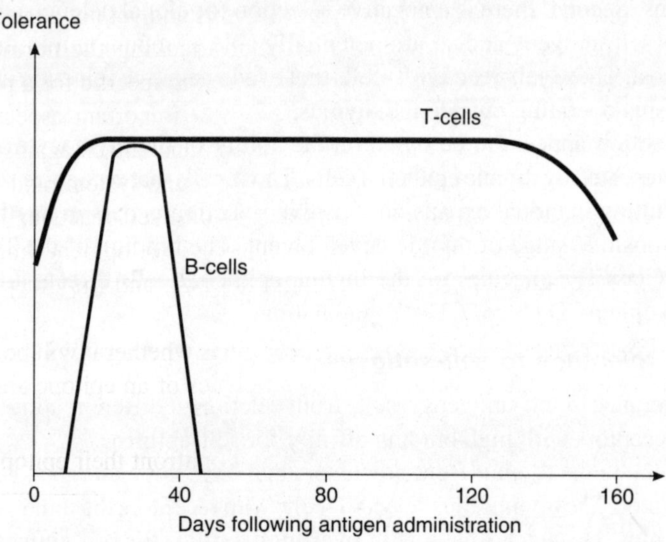

Figure 8.1 The duration of tolerance in T- and B-cells in mice.

Since helper T-cells are required for both cell- and antibody-mediated immune responses, an animal may be functionally tolerant even though its B-cells are fully responsive to an antigen.

There are five possible ways in which self-reactive lymphocytes may be prevented from responding to self-antigens. These are:

1. Clonal deletion, that is physically deleting cells from the repertoire at some stage during their maturation.
2. Clonal abortion, which is prevention of the differentiating of the immature cell.
3. Clonal anergy: down regulating the mechanism of immune response.
4. Suppression: which is inhibiting cellular activity through interaction with other cells.
5. Blockade of antibody forming cells.

Which of these fates awaits the self-reactive lymphocytes depends upon factors like

(i) The stage of maturity of the cell being silenced.
(ii) The affinity of its receptor for the self-antigen.
(iii) The nature of this antigen.
(iv) Concentration of the antigen.
(v) The distribution of this antigen in tissues.
(vi) Its pattern of expression.
(vii) The availability of co-stimulatory signals.

If there is no co-stimulation, antigen is more likely to switch off the cells ability to respond than it is to stimulate an immune response.

8.1.1 T-cell Tolerance

The ability of T-cells to respond to an antigen depends on the availability of the correct antigen receptors. Thus, tolerance to a specific antigen results if an animal lacks T-cells with receptors specific for that antigen.

First, there is a positive selection of T-cells carrying receptors that can recognize the animals own MHC antigens. Second, there is a negative selection (or clonal deletion) of those T-cells that bear receptors for self-antigens and so are potentially able to mount autoimmune responses that may damage the host. Those self-reactive T-cells that evade negative selection can be made tolerant by other mechanisms operating outside the thymus.

Positive selection appears to be based on the ability of the thymocytes to respond to self-MHC molecules presented by thymic epithelial cells. Those cells that recognize self-MHC molecules are stimulated resulting in clonal expansion. Positive selection occurs in the thymic cortex at the immature double-positive stage of thymic development. The binding of the TCR to either MHC Class I or MHC Class II molecules on the thymic epithelial cells directs differentiation of the selected cells into either $CD4^-8^+$ or $CD4^+8^-$ populations.

Central thymic tolerance to self-antigens

Central thymic tolerance to self-antigens results from deletion of differentiating T-cells that express antigen-specific receptors with high binding affinity for self-antigens.

T-cells develop in the thymus from precursors that have not undergone gene re-arrangement of their T-cell receptor. The thymus thus selects T-cells with receptors that bind to antigen associated with MHC molecules, but deletes cells with over-high avidity for self-antigens. There is a high proliferative rate of thymocytes. At the same time there is a massive rate of cell death; the vast majority of double positive thymocytes die within the thymus. The reasons for this can be aberrant TCR gene rearrangement, negative selection and failure to be positively selected.

T-cells that have some degree of binding avidity for polymorphic regions of MHC molecules are selected for survival. This ensures that the selected T-cells will be self-MHC restricted. Positive selection does not prevent the differentiation of T-cells bearing TCR of high binding strength for self-peptides and MHC molecules. Therefore some form of negative selection must operate to silence these strongly self-reactive cells.

The elimination of T-cells capable of responding to self-antigens (clonal deletion) also occurs in the thymus as the T-cells develop, which depends on the affinity of TCR for the self-antigen. T-cells that bind self-antigens with high affinity are deleted earlier and more completely than low affinity cells.

Clonal deletion is the predominant mechanism of negative selection. The timing and localization of negative selection depend on a variety of factors including the

- Accessibility of developing T-cells to self antigens.
- Combined avidity of the TCR and the accessory molecules (CD4 or CD8) for self-MHC self-peptide complex.
- Identity of the deleting cells.

Negative selection is normally a function of the thymic dendritic cells or macrophages, which are predominantly present at the cortico-medullary junction and are rich in Class I and II MHC proteins and therefore can bind the T-cells that have high avidity for self-peptides. Specialized veto cells bearing self-epitopes impart a negative signal, killing the self-reactive clone. Veto signals occur when a T-cell expressing a TCR for self-epitope binds to a veto cell that expresses the self-epitope. Once binding has occurred, the T-cell is killed.

It is not clear how both positive and negative selection can occur in the thymus and not remove all the T-cells. Thus, in one case the self-reactive T-cells are selected and in the other they are destroyed. One hypothesis to explain this contradiction says that although all cells that can bind self-MHC molecules are positively selected, only those that bind MHC molecules with very low or very high affinity are deleted. Thus, moderate affinity clones survive and are able to present foreign antigens.

If the concentration of a specific antigen is high, multiple TCRs will be occupied on each thymocyte. This generates a signal that triggers apoptosis and thus, negative selection. On the other hand low concentrations of antigen occupy only a few TCRs on each thymocyte, causing positive selection and thymocyte proliferation.

Self-tolerance is essential for survival. The body does not rely only on negative selection of thymocytes to ensure that self-reactive T-cells are suppressed. They are suppressed by other mechanisms (Figure 8.2). This functional suppression is called **clonal anergy**, which is the prolonged, antigen-specific suppression of T-cells and depends on the signals delivered to the T-cell. Occupation of the TCR by a MHC-epitope complex in the absence of a co-stimulatory signal leads to clonal anergy. This second signal is normally provided by the binding of CD80 on an antigen presenting cell to CD28 on the T-cell. Once induced, anergy can last for several weeks. Exposing the cell to high levels of IL-2 may reverse the aneroid state. Suppressor T-cells can also suppress inappropriate T-cell responses and give rise to functional tolerance.

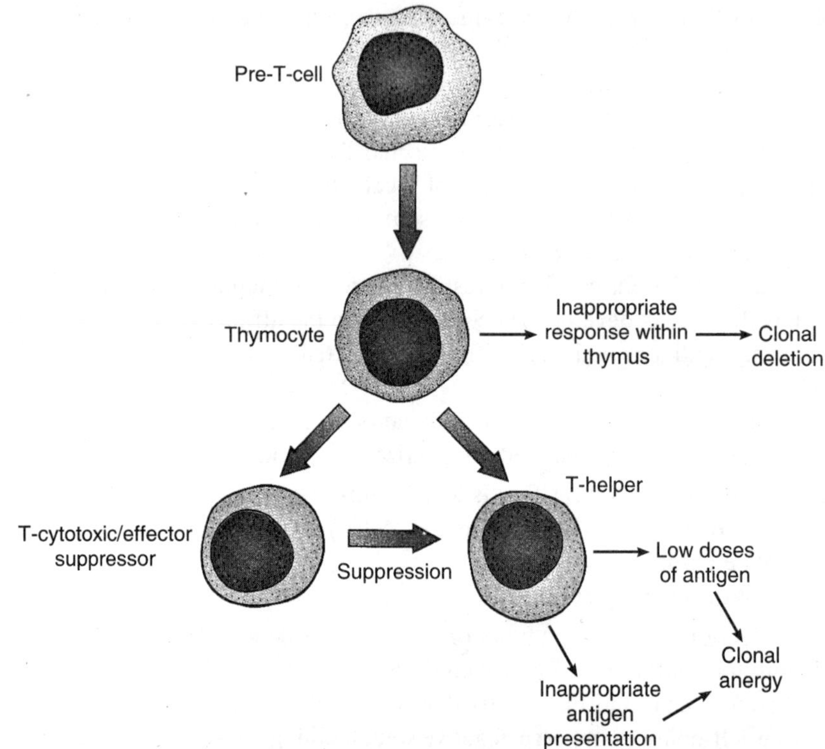

Figure 8.2 Some of the mechanisms of induction of tolerance in T-cells.

Very high doses of antigen induce a form of tolerance known as **immune paralysis.** (Figure 8.3) The high doses of antigen probably bypass antigen-presenting cells, reach helper T-cells directly and make them anergic. T-suppressor cells appear to be involved in tolerance induced both by very low doses of antigen (low dose tolerance) and by de-aggregated antigen.

Superantigens bind to a specific Vβ domain of an α/β TCR. Some superantigens are produced by **retroviruses** that can infect sperm or egg cells and integrate themselves fully into the animal's genome. The proteins made by these integrated viruses are self-antigens.

Post-thymic tolerance to self-antigens

Some self-reactive T-cells escape thymic censorship. Potentially self-reactive T-cells can sometimes ignore their self-antigen. This ignorance of auto antigens occur if self-reactive T-cells are unable to penetrate an endothelial barrier that separates cell bearing self-antigens, or cannot be activated even after crossing these barriers because

(a) The self antigen is present in a very low concentration.
(b) The self antigen is present on tissue cells that express little or no MHC molecules.
(c) The T-cells that perceive the self-antigen require help to be fully activated.

Such ignorance of self-antigens is not a form of immunological tolerance, but is an important factor in learning how some autoimmune diseases arise.

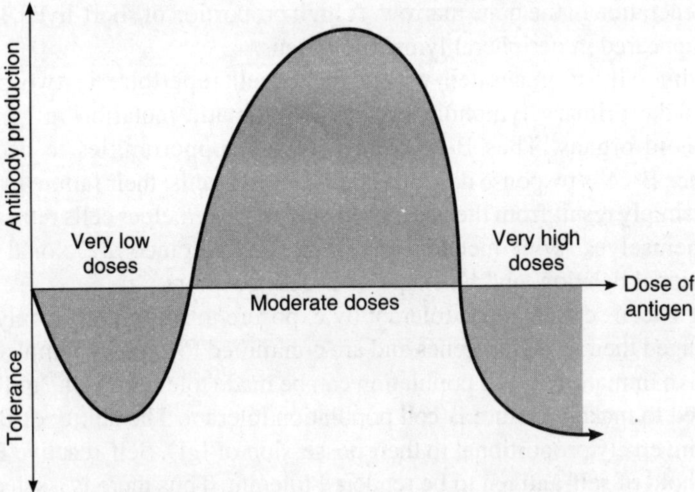

Figure 8.3 High and low dose tolerance.

8.1.2 B-cell Tolerance

High affinity IgG production is T-cell dependent. Moreover the threshold of tolerance for T-cells is lower than for B-cells. Therefore, lack of help from T-cells is the explanation for non-self reactivity by B-cells (Figure 8.4).

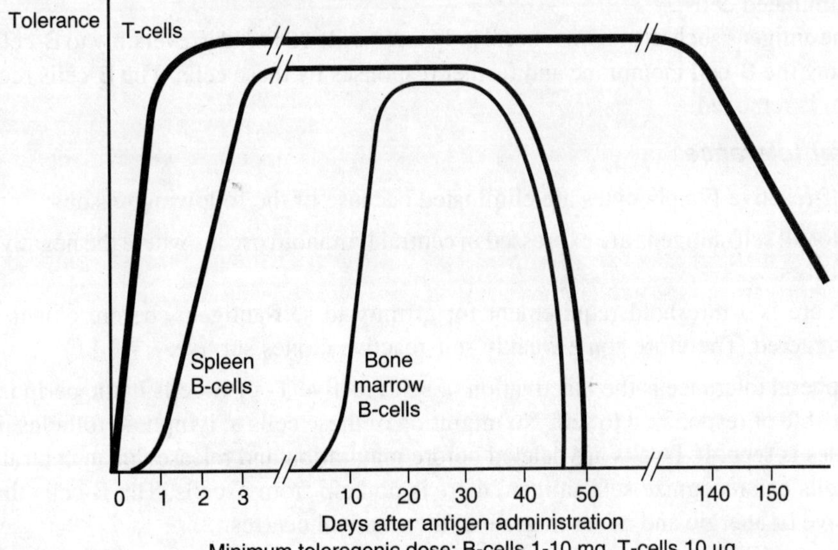

Figure 8.4 The relative susceptibilities of T-cells and B-cells to tolerization in vivo.

Tolerance is also imposed on B-cells. Deletion of self-reactive B-cells takes place in a bone marrow and peripheral lymphoid organs. Differentiating B-cells that express surface immunoglobulin receptors with high binding affinity for self membrane-bound antigens are deleted soon after their generation in the bone marrow. A high proportion of short-lived, low-avidity, auto-reactive B-cells appeared in peripheral lymphoid organs.

Antibody diversity is generated within the B-cell repertoire in two phases—by gene rearrangement in the primary lymphoid organs and somatic mutation in germinal centres in peripheral lymphoid organs. Thus B-cells have several opportunities to become reactive to self-antigens. Since B-cells response depends upon T helper cells, their failure to produce anti-self antibodies could simply result from the absence of self-reactive helper cells rather than any change in the B-cells themselves. Four mechanisms of B-cell tolerance are clonal abortion, clonal exhaustion, functional deletion, and blockage of B-cell receptors.

Developing B-cells can be made tolerant by exposure to self-antigens only when they have successfully arranged their V region genes and are committed to express complete IgM molecules on their surface. An immature B-cell population can be made tolerant by one millionth of the dose of antigen required to make a mature B-cell population tolerant. The ability of B-cells to develop tolerance is also inversely proportional to their possession of IgD. Self-reactive B-cells must bind to a critical threshold of self-antigen to be rendered tolerant. Thus there is a selective silencing of high affinity B-cells. Clonal abortion occurs only if B-cells encounter antigen early in their development within the bone marrow.

B-cells can be stimulated to differentiate into short-lived plasma cells by repeated exhaustive antigenic stimulation. If all B-cells develop into plasma cells, no memory B-cells will remain to respond to subsequent doses of the antigen. This is known as **clonal exhaustion**. Induction of tolerance also involves giving an animal immunosuppressive drug along with an antigen. These drugs may either lower the threshold for tolerance induction or block the differentiation of antigen-stimulated cells.

Some antigens such as pneumococcal polysaccharide can bind irreversibly to B-cell receptors, thus freezing the B-cell membrane and further responses by these cells. The B-cells recover when the antigen is removed.

Peripheral tolerance

Not all self-reactive lymphocytes are eliminated because of the following reasons:

- Not all self-antigens are expressed in central lymphoid organs where the negative selection occurs.
- There is a threshold requirement for affinity to self-antigens, before clonal deletion is triggered. Therefore some weakly self-reactive clones survive.

Peripheral tolerance is the inactivation of self reactive T- or B-cells in the periphery making them incapable of responding to self. No migration of these cells to lymphoid follicles in spleen or lymph nodes is seen. If T-cells are deleted before maturation and release due to central tolerance; and if B cells can recognize self-antigen, there is no help from T-cells. The B-cells thus become unresponsive or anergic and never migrate to the terminal centres.

There are CD28 present on T-cells and B7 present on antigen presenting cells. Interaction between these two provide the necessary co-stimulatory signals required for T-cell activation. CTLA-4 or CD152, expressed on T-cell membrane, is an inhibitory receptor. It binds to B7 and inhibits T-cell activation.

Regulatory T-cells

Peripheral tolerance may be induced by T_{reg}-cells. They act in secondary lymphoid tissues and at sites of inflammation. They down regulate autoimmune processes. T_{reg}-cells are a subset of CD4 T-cells that express increased levels of IL-2Rα chain (CD25). They arise from T-cells that have intermediate affinity for antigens in the thymus. Some of these cells up-regulate the transcription factor Foxp3 and then develop into T_{reg}-cells that can suppress reaction to self-antigens. Suppression is regulated partly through the production of cytokines, including IL-10 and TGFβ.

Antigen sequestration to protect self-antigens. The sequestration or compartmentalization of self-antigens is seen so that they do not encounter reactive lymphocytes under normal circumstances. Breaching of blood brain barrier can cause reaction against components or central nervous system.

Antigens that induce tolerance are called tolerogens rather than immunogens. An antigen presented to T-cells without appropriate co-stimulation results in a form of tolerance known as anergy. Factors promoting tolerance rather than stimulation of the immune system by a given antigen include:

- High doses of antigen
- Persistence of antigen in host
- Intravenous or oral introduction
- Absence of adjuvants
- Low levels of co-stimulators.

Thus, possible mechanisms of tolerance inductions (Figure 8.5) are the following:

Clonal deletion: Cells capable of recognizing a specific antigen have been deleted, or removed. Hence, there is no possibility of initiation of an immune response.

Suppression by T_S-cells: T_S-cells produce T_SF, which blocks growth and differentiation of effector cells. Antigen recognition might occur, but expression of immune response is blocked.

Anti-idiotype antibody: Anti antibody blocks the capability of specific antibody to function; B- or T-cell function is expressed, but the effects are neutralized.

Blocked receptors: These are formed with antigen receptors that prevent functional interactions with intracellular activating systems: effective binding and therefore, initiation of immune responses are prevented.

Lack of contra suppression: This is because of the inability to release the block by T_S-cells on T_H-cells. Recognition may be possible, but clonal expansion and differentiation do not occur.

For maintaining tolerance the deletion of lymphocytes clones that may react with self components is necessary during early maturation. Self-reactive lymphocytes are removed or inactivated after encounter with self-antigen. This was established by the work of Goodnow in 1991. Later David Nemazee and his colleagues demonstrated that some developing B-cells can undergo a process which they termed as **receptor editing**. In cells undergoing receptor editing, the antigen specific V region is edited with a different V region gene segment switched for the autoreactive V gene segment. Receptor editing occurs most often with the V_L rather than the V_H segment. Receptor editing as well as clonal deletion or apoptosis is recognized as one of mechanisms that lead to central tolerance in developing B-cells. By a similar mechanism, T-cells developing in the thymus that have a very high affinity for self-antigen are deleted through apoptosis.

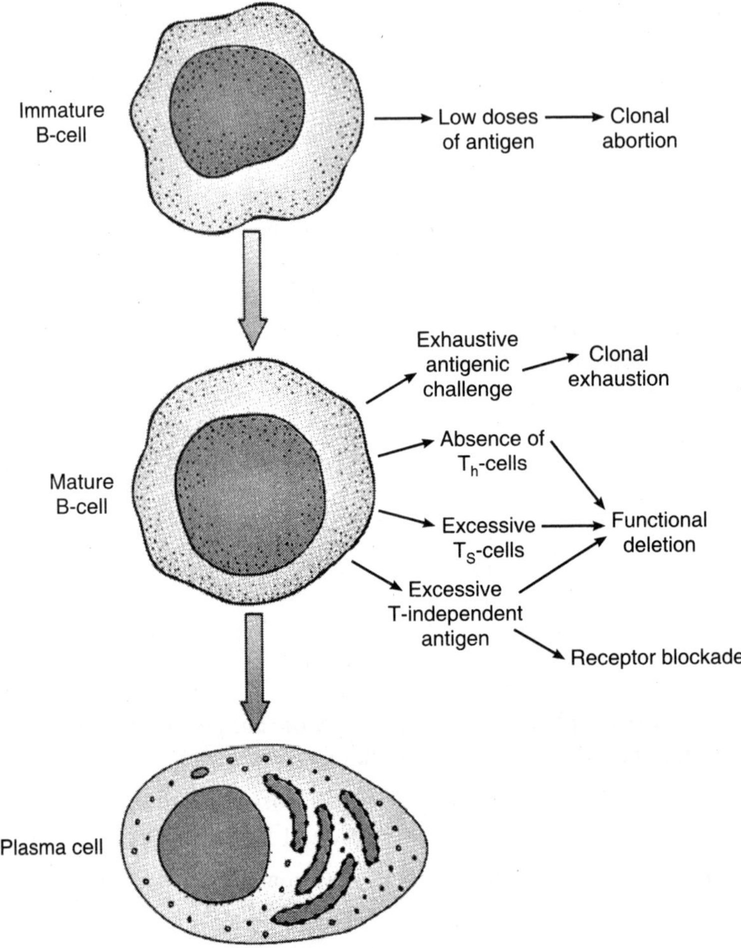

Figure 8.5 The mechanisms of induction of tolerance in B-cells.

8.2 IMMUNOSUPPRESSION

Interference with the induction or expression of the immune responses is called **immunosuppression**. Because of the non-specific nature of immunosuppressive effects of drugs, the patient undergoing this therapy is susceptible to infection; they are also more susceptible to developing cancers. Immunosuppressive therapy is used for patients receiving grafts and those suffering from autoimmune diseases.

Artificial tolerance can be induced by various ways that may be exploited clinically to prevent rejection of foreign grafts and to deal with autoimmune diseases. This can be induced by a number of ways:

1. Antibodies to T-cell co-receptors induce tolerance to transplants.
2. Soluble antigens readily induce tolerance.
3. Oral administration of antigens induces tolerance.

4. By targeting antigens to naive B-cells.
5. By veto and suppressor cells.
6. Antagonistic peptides that signal inappropriately can induce tolerance.
7. Extensive clonal proliferation can lead to exhaustion and tolerance.

Tolerance need not be total. It can be generated in selected segments of the immune system. The duration of tolerance depends on the persistence of the antigen and the ability of the bone marrow to generate fresh antigen-sensitive cells. Treatment that promotes bone marrow activity, like low doses of radiation, hastens the fading of tolerance, whereas immunosuppressive drug treatment has the opposite effect.

8.2.1 Specific Immunosuppression

Antibodies can be used as **specific immunosuppressive agents**. Depending on their specificity, some antibodies can block the formation of other Abs either by hindering the reaction of the immunogen with appropriate cell surface receptors, or by inactivating lymphocytes (specially T-cells).

Suppression can be achieved by making use of Ab as an anti-immunogen, where antibody is added to the immunogen before its injection (or are present in the animal due to previous immunization). The induction of Ab formation is blocked. The inhibition increases when antibody is in excess. The Ab-Ag complexes are rapidly degraded.

Specific immunosuppression can also be achieved by making use of Ab as antireceptor. This is possible because depending upon the specificity and conditions of administration, the anti-Ig suppresses the formation of Ig of the corresponding class (isotype, allotype or idiotype). This is done either by eliminating B-cells with target Ig as surface receptor or by blocking their stimulation by antigen.

A third way to achieve specific immunosuppression is by using Anti-Lymphocyte Serum (ALS). ALS is usually prepared in horse or rabbit against another species thymus cells, white blood cells or cultured lymphocytes. The Anti-Lymphocyte Globulin (ALG) causes preferential inactivation of the long-lived pool of recirculating lymphocytes (which are mainly T-cells). This prolongs the survival of allografts. It also reduces the antibody formation in primary responses to many T dependent antigens, if given just before the antigen. However, if the F_C domain is removed, ALG loses its effectiveness.

8.2.2 Non-specific Immunosuppression

Non-specific immunosuppression can be done with the help of whole body X-irradiation and a number of cytotoxic drugs. In total lymphoid irradiation the bone marrow is shielded and the individual's lymphoid tissue, lungs and other vital non lymphoid tissues are subjected to irradiation. The irradiation induces the formation of large granular lymphocytes, which lack T, B and macrophage markers which non-specifically suppress the cytolytic reaction involved in grafting. X-irradiation and cytotoxic drugs prevent the initiation of Ab formation. However, once antibody formation has started they cannot interrupt unless given in very large doses that are cytotoxic for cells in general. It is seen that these non-specific immunosuppression agents are more effective in blocking primary than secondary response. This may be because either the memory cells are more resistant than virgin antigen sensitive cells, or the increased number of cells that can respond increases the probability that some cells will initiate antibody formation before they are blocked.

The non-specific immunosuppressants may be anti-metabolites, lympholytic agents or both lympholytic and anti-proliferative. The anti-metabolites or anti-proliferatives block cell proliferation. They are specially toxic for dividing cells and suppress antibody formation best when administration is begun two days after antigen induced proliferation is active. They only damage cells with replicating DNA. Examples of anti-metabolites are purine analogues like 6 mercapto purine, pyrimidine analogues like 5 fluoro uracil, and folic acid analog like amethopterin. The anti-metabolites are effective in the inductive phase of Ab responses, which is between immunogen administration and antibody production.

Lympholytic agents block cell division by cross-linking strands of DNA resulting in massive destruction of lymphocytes. These have to be given at frequent intervals for sustained immunosuppression. Examples are alkylating agents and corticosteroids. Corticosteroids cause extensive destruction of small lymphocytes. They also inhibit inflammation and are often used to suppress allergic inflammation. The lympholytic agents are effective when given in the pre-inductive phase, which is before the administration of antigen.

X-rays and radiomimetic alkylating agents act as both lympholytic and anti-proliferative agents. Whole body irradiation with sub-lethal doses of X-rays suppresses immune response to most immunogens for many weeks. Suppression is greatest when irradiation is done 24 to 48 hours before antigen is given, for causing massive disintegration of lymphocytes to occur promptly. After three to four weeks, cell proliferation is resumed and lymph nodes become normal shortly. However, damage to chromosomes can be permanent. It could also impair the capacity of surviving lymphocytes to undergo mitosis.

All immunosuppressive drugs are ineffective in the productive phase of Ab synthesis, unless given in highly toxic doses.

Some anti-microbial drugs also inhibit Ab synthesis. Chloramphenicol and actinomycin D can act as immunosuppressants.

The drugs that are employed for immunosuppression are also used for cancer chemotherapy because these drugs are highly toxic to dividing cells. These anti-mitotic drugs are toxic to the bone marrow cells and highly proliferative intestinal cells. The drugs commonly used for immunosuppression during tissue transplants are:

Azathioprine: It has a preferential effect on T-cell mediated reactions. This is a competitive inhibitor of the enzyme concerned in the synthesis of guanilic acid and adenylic acid. The net result is inhibition of nucleic acid synthesis. Both B-cell and T-cell proliferation is reduced by azathioprine.

Methotrexate: It is an inhibitor folic acid and inhibits nucleic acid synthesis. Folic acid is responsible for the production of dihydrofolate, which is one of the precursors of nucleic acid synthesis.

N-mustard derivative of cyclophosphamide: This attacks the DNA directly by alkylating its nitrogenous bases and cross-linking, thereby preventing the correct duplication during cell division.

These drugs exert their toxic effect on cells during mitosis and therefore, are most effective when administered after the presentation of antigen (graft) when the antigen sensitive cells are dividing.

Cyclosporin A: It is capable of penetrating the antigen sensitive T-cells and selectively blocking the transcription of lymphokine mRNA. The side effects of cyclosporin A include nephrotoxicity.

Steroids like prednisone: Prednisone interferes at various points on the immune response affecting the generation of cytotoxic effector cells. It is also anti-inflammatory.

8.3 APPLICATIONS OF TOLERANCE

Understanding tolerance can promote the tolerance of foreign tissue grafts or to control the damaging immune responses in hypersensitivity states and autoimmune diseases. Treatment with monoclonal non-depleting anti-CD4 and anti-CD8 antibodies has been used successfully for transplanting foreign tissues or organs. Using immunosuppressants for tolerance is very useful for maintaining a tolerant state needed for organ transplantation. A combination of azathioprine (an anti-metabolite), predinisone (a corticosteroid) and ALG is commonly used for immunosuppression before organ transplantation.

However using immunosuppressants pose some complications. These agents are highly toxic for various cells. Latent infections are activated as the person is immunosuppressed. An increased susceptibility to serious infections and an increased incidence of cancers is also seen.

MULTIPLE-CHOICE QUESTIONS

1. What cell types can be made tolerant?
   (a) T-cells alone (b) B-cells alone
   (c) T- and B-cells (d) macrophages
2. The major step in inducing T-cell tolerance occurs in:
   (a) thymus (b) bone marrow
   (c) spleen (d) lymph nodes
3. When a T-cell is made tolerant by receiving inappropriate co-stimulation, it is called:
   (a) negative selection (b) positive selection
   (c) clonal anergy (d) clonal selection
4. Very high doses of antigen induce:
   (a) negative selection (b) clonal selection
   (c) immune paralysis (d) positive selection

REVIEW QUESTIONS

8.1 Discuss the ways in which a foetal animal becomes tolerant to a self-antigen.

8.2 How are thymocytes positively and negatively selected at the same time?

8.3 Explain:

(i) natural immunological tolerance
(ii) mechanism of tolerance
(iii) high and low dose tolerance
(iv) receptor editing

8.4 Write a detailed note on specific and non-specific immunosuppression.

8.5 Discuss the role of regulatory T-cells in tolerance.

8.6 What is central and peripheral tolerance?

9

HYPERSENSITIVITY

9.1 HYPERSENSITIVITY

An immune response evokes the formation of many effector molecules that act to remove the antigen by various mechanisms. Usually these molecules induce a mild inflammatory response that eliminates antigen without damaging the host tissue. Sometimes this inflammatory response can have adverse effects that can lead to tissue damage or even death. This inappropriate immune response is known as **hypersensitivity** and may develop either through humoral or cellular response. Thus, hypersensitivity is an exaggerated response of an organism to a persistent antigen, which may result in inflammation and possible tissue damage.

When an immunologically sensitized individual comes in contact with the same antigen for a second time, it leads to secondary boosting of the immune response. The second response may be excessive leading to tissue damage if the antigen is present in large doses or if the cellular and humoral immune response is at a heightened level. A heightened immune response evokes mainly effective molecules, which generally induce mild subclinical, localized inflammatory reaction that help in the removal of antigen without causing excessive damage to the tissue in the host. Sometimes this inflammatory response may be more severe and cause tissue damage. Such a reaction is called hypersensitivity or allergy. There are several types of hypersensitivity reactions, which can be distinguished on the basis of effector molecules generated during the course of reaction.

Hypersensitivity has been classified as Class I, II, III and IV hypersensitivity. The first three types of hypersensitivity are antibody dependent, while the fourth type is cell mediated, with some overlapping between the various types. The immune reactants and the effector mechanisms of the different types of hypersensitivity reactions are very diverse (Table 9.1).

Table 9.1 Immune Reactants and Effector Mechanisms of Different Types of Hypersensitivity

	Type I	*Type II*		*Type III*		*Type IV*
Immune reactant	IgE	IgG	IgG	IgG	T-cells	T-cells
Antigen	Soluble antigen	Cell or matrix associated antigen	Cell surface receptors	Soluble antigen	Soluble antigen	Cell associated antigen
Effector mechanism	Mast cell activation	Complement, FcR^+ cells	Antibody alters signalling	Complement, phagocytes	Macrophage activation	Cytotoxicity

Certain class of antigens referred to as allergens are responsible for type one hypersensitivity action. These allergens stimulate humoral immune response resulting in the production of antibody secreting plasma cells and memory cells. The plasma cells secrete excessive quantity of IgE. F_C receptors for IgE are present on mast cells and basophilic membrane. On binding to the F_C receptors on the basophils and mast cells, the IgE sensitizes them. When such sensitized cells react with allergen through IgE molecules, they bring about degranulation of the cells.

Most other allergens enter the body through the oral or nasal aperture and bring about allergic reaction on mucus membrane surface, where IgE is available. Many of the allergens are small proteins or protein-bound substances having a molecular weight of 15,000 to 40,000. Some of them are haptens capable of reacting with body proteins and forming a stable complex.

9.1.1 Type I Hypersensitivity

This is also known as immediate hypersensitivity, is mediated by an antibody (IgE) and occurs immediately after a second contact with the same antigen, known as **allergen** (like pollen grains). During the first contact, the stimulating antigen triggers a series of events at the site of antigen entry, resulting in the production of IgE, which binds to Fc receptors on the surface of tissue mast cells and blood basophils and sensitize them. A later exposure of the same allergen cross links the membrane bound IgE on sensitized mast cells and basophils, and causes Ca^{++} to enter into these cells, resulting in the release of cellular granules. Other molecules (anaphylatoxins), mainly products of complement activation, may also induce release of granules containing various mediators. Thus it is the combination of antigen with IgE bound to receptors on the mast cell surface that causes the release of active molecules from mast cells which results in Type I hypersensitivity (Figure 9.1). The released mediators (e.g. prostaglandins, eosinophil chemotactic factors, and histamines) trigger the clinical symptoms observed in an allergic reaction. The effect may either be systemic or localized depending on the extent of mediator release, and is probably because of an imbalance in the helper cell activity. Type I hypersensitivity is characterized by a rapid time course, by edema and inflammation, and by infiltration of affected tissues by eosinophils. The tendency to develop this kind of hypersensitivity is inherited.

Several general factors are known to influence the expression of allergies. More males than females are affected by allergies, and the prevalence of allergy rises to peak at about 20 to 24 years before declining gradually. Breast-feeding tends to prevent allergy. Environmental and genetic factors have also important influence on the development of allergy.

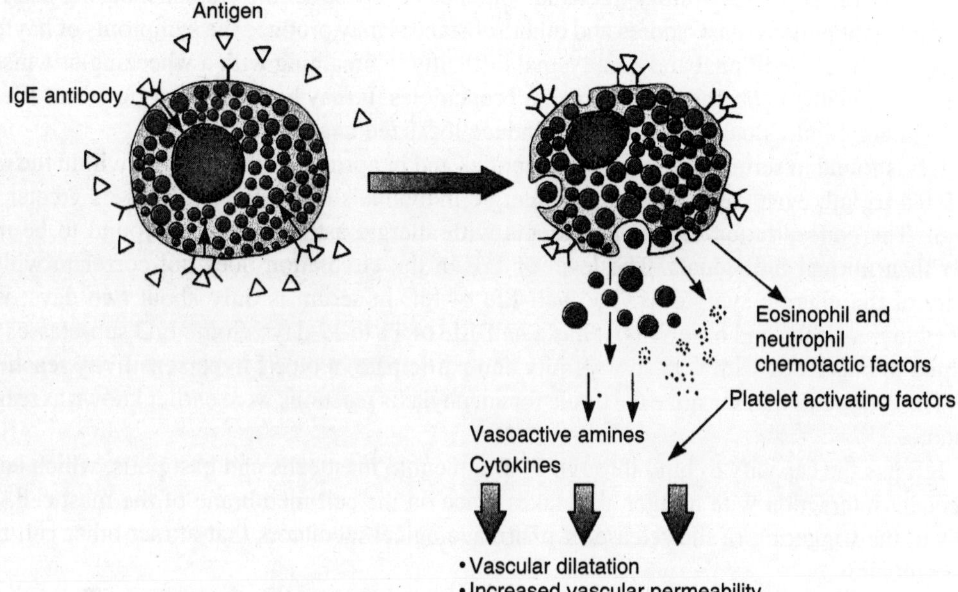

Figure 9.1 The mechanism of Type I hypersensitivity.

The allergic reaction may be either a local inflammation or a generalized anaphylaxis. A localized reaction occurs as an immediate response to mediators released from mast cell degranulation. It is characterized by a cutaneous reaction that appears as a *wheal and erythema reaction*, local edema, and itching at the site of introduction of the particular allergen (e.g. bee sting). The generalized reaction of antigen-specific immune reaction, also known as anaphylaxis, is produced by mediators like cytokines and vasoactive amines (e.g. histamine released from mast cells). The reaction is characterized by distention of blood vessels (vasodilation), smooth muscle constriction, bronchospasm, and edema (accumulation of excessive fluid) that may lead to death.

A number of individuals are susceptible to natural sensitization by a variety of environmental antigens, including pollen, spores, dust, drugs and foods. These individuals are genetically susceptible, a situation known as **atopy**. These hyper active reactions in humans can also be passively transferred to non-hypersensitive individuals by injecting antibodies. To avoid generalized reaction and sensitization, a localized transfer is performed. This procedure is called the Prausnitz–Kustner (PK) reaction. In this procedure, serum from a sensitized individual is injected intradermally into normal recipient. After one or two days, antigen is injected into the same site and a localized wheal and erythema reaction occurs within minutes.

Some individuals inherit a greater tendency to mount an IgE response. Such people are said to be in the state of atopy. This is a hereditary abnormality, and is a predisposition to develop immediate hypersensitivity reactions against common environmental antigens. Allergenicity is a result of a complex series of interactions that involve allergen, its dose, the sensitizing route, an adjuvant and the genetic constitution of the recipient. Atopic individuals make IgE constantly and have unusually high levels of this immunoglobulin.

Atopic allergy

Hayfever is one example of atopic allergy. It is characterized by watery discharge from the mucus membrane of the upper respiratory tract and conjunctiva. These set off violent sneezing and nasal discharge. Plant pollens, dust, spores and other substances may produce the symptoms of hayfever.

Asthma is a conditional and paraxysmal difficulty in breathing with a wheezing or whistling sound produced due to obstruction of smaller bronchioles. It may be either allergic or not allergic.

In allergic individuals, various foods induce localized anaphylaxis too.

IgE is found in serum in very small quantities and in normal individuals fall within the range of 0.1–0.4 μg/ml; even the most severely allergic individuals rarely have IgE levels greater than 1 μg/ml. The concentration of IgE in patients with allergic asthma has been found to be much higher than normal individuals. The level of IgE in the circulation does not correlate with the severity of the allergic symptoms. The half-life of IgE in serum is only about two days; when attached to mast cells and basophils, it has a half-life of 11 to 12 days. Some IgG sub-classes may also bind to mast cells with very low affinity and participate in type I hypersensitivity reactions.

The antibodies that sensitize the tissue for anaphylaxis reactions were earlier known as reaginic antibodies.

IgE has the capacity to bind through F_C fragment to mast cells and basophils, which lead to subsequent interaction with antigen that takes place on the cell membrane of the mast cell. This results in the triggering of the release of pharmacological mediators that attract other cell types like eosinophils.

The B-cell response leading to the production of IgE occurs in two stages. First a resting B-cell switches from IgD/IgM production to IgE production under the influence of IL-4 being produced by T_h2 cells. T_h1 cells produce IFN-γ, which has an opposing effect. The ratio of T_h1 to T_h2 cells engaged in an immune response probably determines the level of IgE produced. The second stage is the development of an IgE producing B-cell into a plasma cell, which is controlled by cell surface protein called **FcεRII** or **CD23** that links the B-cells to dendritic cells, thus preventing their destruction by apoptosis. The FcεRII expression is regulated by the presence of IL-4, which stimulates its production, and by IgE, which suppresses its production.

There are two types of Fc receptors namely, FcεRI and FcεRII which have got different affinity to bind to IgE. The high affinity receptor FcεRI is mainly found on mast cells and basophils, about 10^4 to 10^5 per cell. The low affinity receptor FcεRII is found on NK cells, macrophages, dendritic cells, eosinophils and platelets. About 30% of B-cells also possess FcεRII and this percentage rises in atopic people. The reagnic activity of IgE depends upon its ability to bind to a receptor specific for the F_C region of the ε heavy chain. FcεRII is also known as CD23 and is present in higher levels on the lymphocytes and macrophages of atopic individuals.

When antigen binds to mast cell bound IgE and cross-links two or more FcεRI structures, a series of reactions are triggered that cause the mast cell granules to migrate to the cell surface, to fuse with the cell membrane, and to release their contents into the surrounding tissues. This is done through several signal transduction pathways. Some of these mast cell responses are extremely rapid. Degranulation can occur within a few seconds after antigen binds to antibody on the cell surface. Drugs or chemicals may also degranulate mast cells. Cells that carry FcεRII such as platelets or macrophages may also be activated by antigen through IgE.

The release of mediators induced by the allergens result in increase in vascular permeability and inflammatory response. The primary mediators are produced before degranulation and stored in the form of granules. The most important rapid mediators are histamine, protease, eosinophil chemotactic factor, neutrophil chemotactic factor, and heparin. The secondary mediators are synthesized after target cell activation. The important secondary mediators are leukotrienes, platelet activating factor, prostaglandins, bradykinin and other cytokines.

Mechanism

On bridging the receptors by cross-linking of IgE molecules on mast cells, there is a rapid breakdown of phosphatidyl inositol to inositol triphosphate, followed by generation of diacyl glycerol and an increase in the intracellular free calcium. It is the activation of phospholipase C that generates both IP3 and diacyl glycerol, which in turn activate protein kinase C.

It is only IgE : allergen ratio of 2 : 1 or greater that could induce degranulation. Calcium level reaches a peak within two minutes of FcεRI cross linkage and this increase is due to uptake of extracellular calcium and release of calcium from intracellular stores in the endoplasmic reticulum. Concomitantly, there is a transient increase in membrane bound adenylate cyclase activity with the rapid peak of cAMP. The cAMP dependent protein kinases phosphorylate the granule membrane proteins, changing the permeability of the granules to water and calcium. The granules swell and fuse with the plasma membrane, releasing their contents. The increase in cAMP is transient and is followed by a drop in cAMP levels, which is necessary for degranulation to proceed. When cAMP levels are increased by drugs, the degranulation process is blocked.

Mast cells can constitutively release cytokines such as IL-4, IL-6, TNF-α, and TGF-β. Once stimulated by antigen they initiate the production and release of many biologically active molecules. Vasoactive amines and enzymes are released that cause capillary dilation and increased blood vessel permeability. Synthesis of leukotrienes and prostaglandins also occurs. The release of preformed and newly synthesized molecules causes the characteristic inflammatory lesions of Type I hypersensitivity (Table 9.2).

Table 9.2 Major Mediators Involved in Type I Hypersensitivity

Mediator	*Major Actions*
Preformed mediators	
Histamine	Smooth-muscle contraction, increased vascular permeability, itching, increased exocrine secretion
Serotonin	Vasospasm and smooth-muscle contraction
ECF-A	Eosinophil chemotaxis
Platelet activating factor	Platelet aggregation and secretion of vasoactive factors
NCF-A	Neutrophil chemotaxis
Mediators generated de novo	
Prostaglandins	Very complex; influence vascular and smooth-muscle tone, aggregation, and immune reactivity
Leukotrienes-C and -D	Characteristic smooth-muscle contraction and increased vascular permeability
Leukotriene-B	Neutrophil and eosinophil chemotaxis
Bradykinin	Smooth-muscle contraction and increased vascular permeability
Serotonin	Smooth-muscle contraction and vasospasm

In short, the release of mediators begins with the aggregation of the Fc receptors. When an allergen binds to a mast cell or basophil, a tyrosine kinase activates phospholipase C that leads to the production of diacyl glycerol and inositol phosphate. This results in rapid influx of Ca^{2+} that activates protein kinase C and adenylate cyclase, thereby increasing the intracellular cAMP levels. The cAMP activates another protein kinase that phosphorylates myosin in intracellular filaments on granules. The mediator rich granules then migrate to the cell's surface and there is exocytosis.

The primary mediators of Type I reactions are produced before degranulation and are stored in the granules. These are histamine, proteases, eosinophil chemotactic factor, neutrophil chemotactic factor, and heparin. Secondary mediators are either synthesized after target-cell activation or are released by the breakdown of membrane phospholipids. These include platelet activating factor, leukotrienes, prostaglandins, bradykinin, and various cytokines. Mast cell granules contain high concentrations of histamine and serotonin (Table 9.3).

Table 9.3 The Constitutive and Inducible Mediators of Hypersensitivity

S. No.	*Mediators*	*Biological Effects*
	Stored in mast cells and basophils	
1.	Histamine	Vasodilation, change in vasopermeability, constriction of smooth muscles, bronchospasm
2.	Proteases (tryptase, chymase)	Role in IgE mediated hypersensitivity not known
3.	Heparin (mast cells)	Complexes with proteases and histamine
4.	Chondroitin sulphate (basophils)	Complexes with proteases and histamine
5.	Eosinophil chemotactic factors	Eosinophil chemotaxis
6.	Neutrophil chemotactic factors	Neutrophil chemotaxis
	Synthesized in activated mast cells and various other cells	
1.	Cysteinyl-leukotrienes (LT-C4, LT-D4, LT-E4)	Change in vasopermeability, bronchospasm
2.	Prostaglandin D2	Change in vasopermeability, bronchospasm
3.	Platelet activating factors	Neutrophil aggregation, aggregation of platelets

Binding of histamine to its receptors induces contraction of intestinal and bronchial smooth muscles, increased permeability of venules, increased mucus secretion and stimulation of exocrine glands. The leukotrienes and prostaglandins are formed after the mast cell degranulation and enzymatic breakdown of phospholipids. The leukotrienes mediate bronchoconstriction, increased vascular permeability, and mucus production. Prostaglandin D2 causes bronchoconstriction. The leukotrienes are 1000 times more potent for bronchoconstriction than histamine is.

Mast cells also release cytokines like IL-4, IL-5, IL-6 and TNF-α, which alter the local microenvironment leading to recruitment of inflammatory cells like neutrophils and eosinophils. Because of the diverse biological effects of the cytokines, mast cells and basophils contribute to a large variety of physiological, immunological and pathological processes.

Proteases present inside the granules of the mast cells destroy nearby cells and activate complement components C3 and C5 to generate anaphylatoxins. Kinins like bradykinin and anaphylatoxins are powerful vasoactive agents. Platelet Aggregating Factor (PAF) makes the platelets to aggregate and release their vasoactive molecules (especially serotonin), and synthesize thromboxanes. PAFs promote neutrophil aggregation, degranulation, chemotaxis, and release of oxygen radicals. They are also major activators of eosinophils. Mast cell granules also release two preformed tetra peptides that are chemotactic for eosinophils. A neutrophil chemotactic factor called neutrophil immobilization factor is also released, which stops the emigration of neutrophils once they have been attracted to an area of inflammation by the chemotactic molecule.

The clinical signs of type one hypersensitivity relate to the release of vasoactive molecules from mast cells and basophils. If the rate of release of active molecules is greater than the body's ability to respond to the rapid changes in its vascular system, the patient suffers from acute anaphylaxis or **anaphylactic shock** and may die. In humans anaphylaxis may follow the rapid administration of a drug or other allergen to which an individual is allergic.

The clinical manifestations of Type I reactions range from serious life-threatening conditions like systemic anaphylaxis, to hay fever and eczema. Systemic anaphylaxis is a shock-like and often fatal state that occurs within minutes of a Type I hypersensitive reaction. Within one minute the animal becomes restless, its respiration becomes laboured, and its blood pressure drops. There is a constriction of smooth muscles of the gastrointestinal tract and the bladder. Bronchiole constriction results in death within 2–4 minutes. The mediators released in the course of the reaction bring on these events (Table 9.3). Massive edema, shock and bronchiole constriction are the major causes of death. Many antigens that trigger this reaction in hypersensitive humans include venom from bee, wasp, hornet and ant stings; drugs like penicillin, insulin and antitoxins; and seafood and nuts (Table 9.4). Epinephrine counteracts the effects of histamine and the leukotrienes by relaxing smooth muscles and reducing vascular permeability. It also improves cardiac output that is necessary to prevent vascular collapse during an anaphylaxis reaction. Epinephrine increases cAMP levels in the mast cells, thereby blocking further degranulation.

Table 9.4 Major Allergens of Humans

Foods	*Proteins*	*Insects*	*Drugs*
Nuts	Mold spores	Ant bites	Salicylates
Chocolate	Animal danders	Dust mites	Sulfonamides
Eggs	Pollens	Bee and wasp stings	Penicillin
Seafood	Vaccines		Local anesthetics
Dairy products	Foreign serum		
Beans, peas			

A common manifestation of localized anaphylaxis is asthma. Airborne or blood borne allergens like pollens, dust, fumes, insect products, or viral antigens trigger an asthmatic attack. This is triggered by degranulation of mast cells and the reaction develops in the lower respiratory tract. This results in contraction of the bronchial smooth muscles leading to bronchoconstriction.

Skin testing identifies Type I hypersensitivity. The potential allergens are introduced at specific skin sites in small amounts. If a person is allergic to the allergen, local mast cells degranulate and the release of histamine and other mediators produce a wheal and flare within 30 min.

Systemic or generalized anaphylaxis

It is a shock which often turns fatal, whose onset takes place within minutes of Type I hypersensitive reaction. The manifestation of generalized anaphylaxis varies with different species in which different organs maybe affected. A few minutes after challenge with antigen the animal begins to scratch, sneeze, cough and may convulse and can collapse and die. This is the cause of respiratory impairment due to constriction of smooth muscles in the bronchioles that leads to bronchial oedema. The sharp decrease in blood pressure and an increase in vascular permeability is also seen. In humans, generalized anaphylaxis occurs with itching, erythema, vomiting, abdominal cramps, diarrhoea and respiratory distress; in severe cases death may also occur.

9.1.2 Type II Hypersensitivity

Type II hypersensitive reactions involve antibody mediated destruction of cells.

In Type II hypersensitivity, destruction of target cell expressing certain antigens on their surface occurs by a process of antibody-dependent cytotoxicity (cell lysis), which may be either complement-mediated or cell-mediated. Antibodies can mediate cell destruction by activating the complement system and by Antibody-Dependent Cell-mediated Cytotoxicity (ADCC).

Type II hypersensitivity results from the destruction of cells by antibodies or complement. An example of this is the destruction of foreign red blood cells as a result of an incompatible blood transfusion. The tissue or cell damage is the direct result of the actions of antibody and complement. Antibodies directed against antigens on the surface of nucleated cells or red blood cells will cause their lysis in the presence of complement.

Typical symptoms include fever, chills, nausea, clotting within blood vessels, pain in the lower back, and haemoglobin in the urine. Delayed haemolytic transfusion occurs in individuals who have received repeated transfusions of ABO-compatible blood that is incompatible for other blood group antigens. This type of reaction is observed in blood transfusion reactions in which host antibodies react with foreign antigens present in the incompatible transfused blood cells. The reaction mediates destruction of the transfused cells. Antibodies like IgE and IgM can mediate the destruction by activating complement system which destroys the foreign cells. Cell destruction can also be brought about by antibody-dependent cell-mediated cytotoxicity.

The reactions develop between two and six days after transfusion of blood. The transfused blood induces clonal selection and production of IgG against a variety of blood group membrane antigens, and symptoms include fever, low haemoglobin, increased bilirubin, mild jaundice and anemia. Free haemoglobin is usually not detected in the plasma or urine in these reactions because RBC destruction occurs in the extra vascular sites.

Certain antibiotics like penicillin, cephalosporin, and streptomycin can adsorb nonspecifically to proteins on RBC membranes forming a complex. Such drug-protein complexes induce formation of antibodies, which then bind to the adsorbed drug on the red blood cells, inducing complement mediated lysis and therefore, progressive anemia. When the drug is withdrawn, the haemolytic anaemia disappears.

Type II hypersensitivity is also seen in haemolytic disease of the newborn or erythroblastosis fetalis. This occurs when there is Rh factor incompatibility between the mother and the foetus. When the mother is Rh negative and carries a Rh positive foetus as a first pregnancy, there is mixing of blood between the two at the time of delivery. The foetal RBC activates the B-cells of the mother and results in the production of IgE and IgM and memory B-cells. Activation of the memory cells occurs in a subsequent Rh positive pregnancy which results in the formation of IgG anti-Rh antibodies. IgG can cross the placental barrier and damage the foetal RBC. This results in anemia, sometimes leading to death of the foetus. It is possible to prevent erythroblastosis fetalis by administering antibodies to the mother against Rh antigen within 24 to 48 hours after the first delivery. These antibodies are called **rhogam** and bind to foetal RBC that enter the mother's blood circulation and destroy them before the B-cells recognize them and get activated and produce memory cells. In such cases, during the subsequent pregnancies the mother is unlikely to produce IgG anti-Rh antibodies and the damage to foetus is prevented.

In antibody-dependent complement-mediated cytotoxicity, cytolysis occurs through a series of reactions that occur when antibody (IgM and some classes of IgG) bind with cell surface antigen and activate complement. The complement fragments (C3b) thus generated participate in the reaction by binding to the target cell membrane and assemble to make a Membrane Attack Complex (MAC), which inserts into a target cell membrane, causing cell lysis. The other complement fragments produced (C3a and C5a) attract neutrophils and macrophages to the reaction site and stimulate mast cells and basophils to produce more chemotactic molecules.

Antibody-dependent cell-mediated cytotoxicity depends on initial binding of specific antibodies to target cell surface antigens. NK-cells and other cells bearing Fc receptors lyse the antibody-coated cells. Target cells are destroyed by the release of cytotoxic substances by these effector cells. Some examples of Type II hypersensitivity are blood transfusion reaction, haemolytic disease of the newborn and Goodpasture's syndrome.

9.1.3 Type III Hypersensitivity

Normally when antigen and antibodies interact they form immune complexes, which are cleared from the body by phagocytic cells. In some cases, large amounts of immune complexes are found which cannot be easily cleared by phagocytic cells and result in tissue damage.

Type III hypersensitivity occurs when immune complexes are not effectively removed by phagocytes and are deposited in various tissues and organs (e.g. kidney, joints, skin, lungs). Type III hypersensitivity occurs as a result of the deposition of complement activating immune complexes in tissues resulting in attracting neutrophils. The lysosomal enzymes of neutrophils and oxidizing radicals cause tissue damage as they attempt to phagocytose the immune complexes (Figure 9.2).

The reaction depends on the quantity of immune complexes as well as their distribution within the body. Type III hypersensitive reactions develop when immune complexes activate the complement system's array of immune effector molecules. These are the C3a, C4a, and C5a complement split products, which are anaphylatoxins that cause localized mast cell degranulation and consequent increase in vascular permeability. They are also chemotactic factors for neutrophils, which can accumulate in large numbers at the site of immune complex deposition. Neutrophils produce lytic enzymes that dissolve the phagocytized immune complexes.

Larger molecular weight immune complexes get deposited in the basement membrane of glomeruli of kidney and the walls of the blood vessels. Small complexes may pass through these membranes and get deposited in the sub-epithelium. Therefore, the type of lesion caused depends upon the site of deposition of immune complexes.

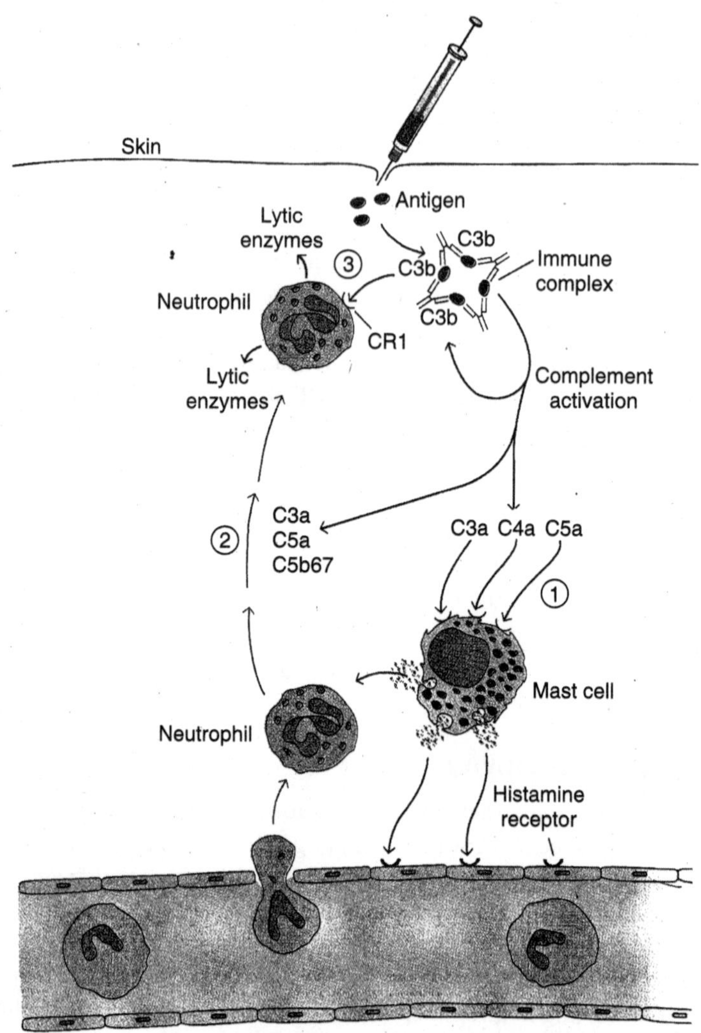

Figure 9.2 Development of Arthus reaction (Type III hypersensitivity). Immune complexes activate complement pathway and produce complement intermediates that (i) cause degranulation of mast cells, (ii) attract neutrophils, and (iii) stimulate release of lytic enzymes from neutrophils.

Immune complexes deposited in the skin, lungs, joints, or walls of blood vessels therefore cause local inflammatory responses. Tissue damage occurs at the site of immune complex deposition. Immune complex formation may occur because of autoimmune disease (rheumatoid arthritis), persistent infection (viral hepatitis) and repeated inhalation of antigenic material. Deposition of immune complexes is promoted by increased vascular permeability in the presence of vasoactive amines like histamine, high blood pressure, and the size and type of antibody in the complex.

The deposited immune complexes activate the complement system, which generates anaphylatoxins, like C3a, that stimulate mast cells and basophils to release chemotactic factors and various amines like histamine. These factors increase vascular permeability and allow effector cells like neutrophils to infiltrate the site of deposition, leading to a local inflammation reaction. Two major types of Type III hypersensitive reactions are seen. One is a local reaction known as the **Arthus reaction**, which results when immune complexes are deposited in tissues. In this reaction there is local necrosis and abscess formation; inflammation and necrosis is also seen in lung, heart, kidney, testes, brain and joints.

If an insect bites a sensitive individual, a localized type one reaction occurs at the site. After 4 to 8 hours a typical Arthus reaction also develops at the site with pronounced erythema and edema.

A second form of Type III hypersensitivity reaction results when large quantities of immune complexes are formed in the bloodstream. If large amounts of antigen enters into the bloodstream and bind to antibody, small immune complexes form. These are not easily cleared by phagocytic cells and can cause tissue damaging Type III reactions at various sites. Within days or weeks after exposure to foreign serum antigens, an individual shows a combination of symptoms that are called **serum sickness**. Serum sickness is characterized by general swelling of lymph nodes and is accompanied by itching, erythematoses, eruption and edema of eyelids, face and ankles. Fever and pain in the joints are seen in severe cases. The symptoms may continue from two days to two weeks. These symptoms include fever, weakness, generalized vasculitis, edema and erythema, lymphoadenopathy, arthritis and sometimes glomerulonephritis. The manifestations of serum sickness depend upon the quantity of immune complexes formed as well as the size which determine the site of the deposition. The antigen antibody complexes get deposited in blood vessels, polymorphonuclear leukocytes also accumulate and severe arthritis follows.

Complexes of antibody with various bacterial, viral, and parasitic antigens have been shown to induce a variety of Type III hypersensitive reactions, including skin rashes, arthritic symptoms and are glomerulonephritis.

9.1.4 Type IV Hypersensitivity

Type IV hypersensitive reactions develop when antigen activates sensitized T_{DTH}-cells which results in the secretion of various cytokines, including interleukin 2 (IL-2), Interferon Gamma (IFN-γ), Granulocyte-Macrophage Colony-Stimulating Factor (GM-CSF), Macrophage Inhibition Factor (MIF), and Tumour Necrosis Factor β (TNF-β).

Type IV hypersensitivity is also known as **Delayed Type Hypersensitivity (DTH)**. DTH is mediated by T-lymphocytes, which had a previous exposure to particular antigen. Following a subsequent contact with the same antigen, these antigen specific T-lymphocytes release cytokines. The released cytokines induce an inflammatory reaction and activate and attract macrophages, causing cellular infiltration, resulting in increased phagocytic activity and increased concentrations of lytic enzymes for more effective killing. It occurs when an antigen that remains within a macrophage after phagocytosis is encountered by previously activated T-cells a subsequent time. The consequences depend upon whether the T-cell is a $CD4^+$ or $CD8^+$ cell.

$CD4^+$ cells secrete lymphokines like IFN-γ, which stimulate the macrophages to release substances that cause inflammation and destruction of bacteria and tumour cells, nonspecifically, regardless of their antigens. The activated macrophages are larger, express more MHC-II proteins and have a number of lysosomal enzymes. They also secrete IL-1, TNF-α, and reactive oxygen intermediates. IL-1 and TNF-α effect breakdown of muscle protein. The reactive oxygen intermediates (superoxide anion, hydroxyl radical and hydrogen peroxide) can damage proteins, lipids, DNA and cell membrane, and are responsible for the destruction of the phagocytosed bacteria and viruses within endosomes.

$CD8^+$ cells react more specifically. They destroy only those cells whose surface antigens they recognize. The activated $CD8^+$ cells act as Cytotoxic T-Lymphocytes (CTLs) and destroy only those cells whose surface antigen MHC I complexes they recognize. When a CTL comes in contact with a cell, it explores it for an antigen. If no antigen is recognized the CTL moves on. However, on recognizing an antigen it becomes activated and releases cytotoxic components by which the recognized cell is lysed.

There are two changes that are seen in target cells when they are lysed by CTLs. A highly hydrophobic protein, perforin, inserts into the target cell membrane and creates pores that act as nonspecific ion channels. Perforin enters the target cells very rapidly and therefore, its diffusion is very limited, thereby sparing the bystander cells. The intracellular calcium ion concentration increases rapidly to about tenfold the normal level, which is toxic for the cell. Secondly, the endogenous endonucleases are activated, resulting in the breakdown of DNA.

The reaction involves a major influx of macrophages and other non-specific inflammatory cells at the site of the reaction. In some cases DTH response causes severe tissue damage. The major cytokines that are involved in DTH reactions are:

(i) IL-2 which functions as an autocrine factor for TD-cells and help in their proliferation.
(ii) IL-3 and GM-CSF which induce local haematopoiesis.
(iii) IFN-γ and TNF-β.
(iv) Activated TD-cells also produce Monocyte Chemotactic and Activating Factor (MCAF) and Migration Inhibition Factor (MIF) which are responsible for drawing the macrophages to the site of DTH reaction and retaining them at the same place.

The macrophages accumulated on the site of such reactions secrete lytic enzymes which destroy the cells that harbour the intracellular pathogens.

DTH reaction does not require participation of complement or antibody. These reactions take about 48–72 hours to develop. Thus a Type IV reaction is characterized by the delay in time and the recruitment of macrophages, which are the major component of infiltrate that surrounds the site of inflammation. Examples of Type IV hypersensitivity reactions are granulomatous skin lesions with *Mycobacterium leprae* and lung cavitation with *Mycobacterium tuberculosis*.

MULTIPLE-CHOICE QUESTIONS

1. Mast cell granules contain:
(a) complement (b) epinephrine
(c) histamine (d) acetylcholine

2. The IgE mediated degranulation of mast cells does not involve:
(a) complement activation
(b) rise in intracellular calcium
(c) synthesis of leukotrienes
(d) release of vasoactive agents from granules

3. On degranulation the mast cells release one of the following:
(a) anaphylatoxins (b) interleukin-1
(c) interleukin-4 (d) immunoglobulin E
4. The Arthus reaction results from local:
(a) red cell lysis (b) mast cell degranulation
(c) cytokine release (d) complement activation
5. Immune complexes cause hypersensitivity by stimulating:
(a) T-cells (b) neutrophil invasion
(c) eosinophil invasion (d) basophil immigration
6. Allergic contact dermatitis can be diagnosed by:
(a) patch test (b) complement fixation test
(c) provocation test (d) intradermal skin test
7. Haemolytic disease of newborn infants can be prevented by:
(a) exchange transfusion (b) administration of allergen
(c) antihistamines (d) administration of anti-Rh antibodies

REVIEW QUESTIONS

9.1 What is hypersensitivity? What are the advantages of Type I hypersensitivity?

9.2 Write notes on:
(i) Atopy
(ii) P. K. reaction
(iii) Systemic anaphylaxis
(iv) Local anaphylaxis

9.3 What is Type IV hypersensitivity? How does it occur?

10

TRANSPLANTATION AND IMMUNOSUPPRESSION

10.1 TRANSPLANTATION

The process of transferring cells, tissues, or organs from one site to another is called **transplantation**. Many diseases can be cured by implantation of a healthy organ, tissue, or cells (a graft) from one individual (the donor) to another in need of the transplant (the recipient).

Tissue may be surgically grafted between different parts of an animal's body. This type of graft is called an **autograft. Isografts** are tissues grafted between genetically identical individuals, like monozygotic twins. **Allografts** are grafts transplanted between genetically different members of the same species. Most of the grafts performed on humans are of this category, as organs are transplanted from a donor who is usually unrelated to the graft recipient. **Xenografts** are grafts between animals of different species. Xenografted organs are very different from the organs of the host and provoke an intense immune response that is very difficult to suppress.

Transplantation is also used to cure diseases. Liver transplants treat congenital defects and damage from viral and chemical agents. Transplantation of pancreatic cells offers a cure for diabetes mellitus. Skin grafts are used to treat burn victims.

There is always a critical shortage of organs available for transplantation, which may be solved in future by using organs from other species. The favoured animal for xenotransplantation is the pig because of the physiological and anatomical compatibility between pigs and humans, and the ability to breed in large numbers of animals rapidly.

10.1.1 Transplantation Antigens

Tissues that are antigenically similar are said to be **histocompatible** and do not induce tissue rejection. Tissues that display significant antigen differences are **histoincompatible** and they induce an immune response that leads to tissue rejection. The antigens determining histocompatibility are encoded by more than 40 different loci. The loci responsible for the most vigorous allograft rejection reactions are located within the **Major Histocompatibility Complex (MHC)**, which is called the HLA complex in humans. Since the MHC loci are closely linked, they are inherited as a complete set, called a **haplotype**, from each parent. Thus, with parent-to-child grafts, the donor and recipient will always have one haplotype in common, but will be mismatched for the haplotype inherited from the other parent.

The expression of MHC on cells is controlled by cytokines. Interferon-γ(IFN-γ) and Tumour Necrosis Factor (TNF) are powerful inducers of MHC expression on many cell types, which would otherwise express MHC molecules only weakly. This is important in graft rejection.

Differences in blood group and major histocompatibility antigens are responsible for the most intense graft rejection reactions. Matching or mismatching of the MHC Class I antigens has a lesser effect on graft survival unless there also is mismatching of the MHC Class II antigens. Successful transplantation even between HLA-identical individuals requires some degree of immunosuppression.

10.1.2 Rejection of Transplanted Tissue

Rejection of transplanted tissue occurs because the immune system of the recipient recognizes and responds to foreign (tissue) histocompatibility antigens expressed on the graft. The host cells respond to the peptide epitopes that are present in the groove of the MHC molecules of the grafted cells.

If an allograft is transplanted, it survives for about a week and is eventually rejected. This is the first-set reaction in which the information about the antigens of the graft reaches antigen sensitive cells, which are usually present in the draining lymph node; and then effector cells from this node invade the graft and mediate its destruction.

If a second graft is transplanted from the same donor, the recipient will reject it in a day or two. The accelerated reaction to the second graft is known as a **second-set reaction**. This second-set reaction is specific for a graft from the original donor and not restricted to a particular site or to any specific organ as MHC molecules are present on all nucleated cells.

In an allograft the second contact with antigen represents a more explosive response than the first and the rejection of the second graft is accelerated. The initial stage shows very poor or no vascularization. The graft area has an increased population of lymphocytes, which include plasma cells. Thrombosis and acute cell destruction is seen by three to four days. Only those allografts derived from the original donor or a related strain show a second set reaction of rejection of graft. The lymphoid cells are stimulated and retained in the memory cells of the first contact with graft antigens. After rejection the host generally produces antibodies against the tissue antigens of the donor graft and it recognizes the same donor immediately on second transplant and causes an accelerated rejection reaction. The lymphoid cells play a primary role in the first set of rejection. T-cell populations CD4 and CD8 are both involved in the graft rejection. Involvement of humoral antibody in the destruction of lymphocytes through cytotoxic Type II reaction in allogeneic grafts is also seen.

The time sequence of allograft rejection varies according to the tissue involved. Generally skin grafts are rejected faster than other tissues such as kidney or heart. The immune response culminating in graft rejection always displays the attributes of specificity and memory.

The first sign of rejection is an accumulation of neutrophils around the blood vessels at the base of the graft. This is followed by an infiltration of **mononuclear cells** (lymphocytes and macrophages) that eventually extends throughout the grafted skin. The whole organ gradually becomes infiltrated with mononuclear cells that cause progressive damage to the endothelial lining of the small blood vessels. Thrombosis of these vessels results in tissue destruction, stoppage of blood flow, haemorrhage, and death of the grafted organ. In second set reaction, blood vessels do not have time to grow into a skin graft, since extensive mononuclear cell and neutrophil infiltration rapidly develops in the graft, leading to destruction of graft.

10.1.3 Mechanism of Graft Rejection

The mechanism of graft rejection entails mainly through a cell-mediated immune response to alloantigens expressed on cells of the graft. The process of graft rejection can be divided into two stages: (i) a sensitization phase, in which antigen-reactive lymphocytes of the recipient proliferate in response to alloantigens on the graft, and (ii) effector stage, in which immune destruction of the graft takes place.

To reach the antigen sensitive cells of the recipient, the graft cells may release soluble MHC molecules that may enter into the host lymphoid tissues. The circulating lymphocytes also encounter foreign MHC molecules either naturally on the cell surfaces or on presentation by host dendritic cells that invade the graft. If the grafted organ contains donor macrophages, dendritic cells or B-cells, then the circulating CD4$^+$ T-cells will also encounter foreign MHC Class II molecules. These CD4$^+$ T-cells move to the draining lymph node and activate CD8$^+$ T-effector cells within that node by secreting interleukin-2. The CD8$^+$ T-cells then leave the node to reach the graft through the blood. On entering the graft, these cells recognize its MHC Class I molecules, and bind and destroy the vascular endothelium and other accessible cells by direct cytotoxicity. This results in haemorrhage, platelet aggregation, thrombosis, and stoppage of blood flow. The grafted organ dies because of failure in blood supply.

During the **sensitization phase**, CD4$^+$ and CD8$^+$ T-cells recognize both major and minor histocompatibility alloantigens. A host T_h-cell becomes activated when it interacts with an Antigen-Presenting Cell (APC) that both expresses an appropriate antigenic ligand-MHC molecule contact and provides the requisite co-stimulatory signal. A population of donor APCs called passenger leucocytes migrate from the graft to the regional lymph nodes. Passenger leucocytes are dendritic cells expressing high levels of Class II MHC molecules of the donor graft; hence, they are recognized as foreign and therefore, stimulate immune activation of T-lymphocytes in the lymph nodes. In the **effector phase**, a variety of effector mechanisms participate in allograft rejection. The most common are cell-mediated reactions involving the delayed type hypersensitivity and CTL-mediated cytotoxicity. Thus, both CD4$^+$ and CD8$^+$ T-cells participate in rejection and the collaboration of both sub-populations results in more pronounced graft rejection.

Since the expression of Class I and Class II MHC molecules is increased in the transplanted tissue during allograft rejection, the graft becomes a very attractive target for the host's cytotoxic T-cells. Antibodies also play a significant role in graft rejection. Antibodies directed against graft

MHC I molecules act with complement and neutrophils, or through antibody-dependent cytotoxic cell activity, to cause vascular endothelial cell destruction.

The less common mechanisms are antibody-plus-complement lysis and destruction by Antibody-Dependent Cell-mediated Cytotoxicity (ADCC). The characteristic of cell mediated graft rejection is an influx of T-cells and macrophages into the graft. Recognition of foreign Class I alloantigens on the graft by host $CD8^+$ cells can lead to CTL-mediated killing. In some cases, graft rejection is mediated by $CD4^+$ T-cells that function as Class II MHC-restricted cytotoxic cells.

TNF-β shows a direct cytotoxic effect on the cells of the graft. A number of cytokines promote graft rejection by inducing expression of Class I or Class II MHC molecules on graft cells. The interferons (α, β and γ), TNF-α, and TNF-β all increase Class I MHC expression, and IFN-γ increases Class II MHC expression as well.

10.1.4 Graft Versus Host Disease

If healthy lymphocytes are injected into the skin of an allogeneic recipient, local inflammation occurs as a result of the lymphocytes attacking the host's cells. In other words the graft attacks the host. If the recipient has a functioning immune system, this **Graft Versus Host (GVH) reaction** is not serious as the recipient is able to destroy the foreign lymphocytes and thus terminate the reaction. If the recipient is unable to reject the grafted lymphocytes then these cells may cause uncontrolled destruction of the host's tissues, leading to disease and eventually death.

GVH may also occur if the recipient of the lymphoid cell graft, has been immunosuppressed or is immunodeficient. This is very important for people who receive bone marrow or thymus allografts to cure leukaemia or congenital immunodeficiencies.

Graft Versus Host Disease (GVHD) is the major complication of bone marrow transplantation, causing severe damage, particularly to the skin and intestine, and is avoided by careful typing, removal of mature T-cells from the graft and the use of immunosuppressive drugs.

The donor T_h-cells from the bone marrow recognize the recipient antigen-MHC complexes displayed on the APCs. The donor T_h-cell gets activated due to antigen presentation and co-stimulatory signal by APC and starts producing IL-2 and undergoes proliferation. The cytokines produced by activated donor T_h-cells induces activation of a variety of effector cells including NK-cells, CTLs and macrophages, which bring about effector phase of GVH disease.

T_h-cells are activated by APCs derived from bone marrow and carrying the MHC Class II molecules. The APCs that activate rejection can come from either the donor or the recipient. Those of donor origin are present in the graft as *passenger leukocytes* (interstitial dendritic cells) and they cause *direct* activation of the recipient's T_h- cells. Thus, passenger cells have a strong influence on graft survival.

In addition to the role of $CD4^+$ T_h-cells, a multiplicity of immunological mechanisms including lymphokines are involved in the process of rejection. The most important lymphokines in cellular rejection are interleukin-2 (IL-2), which is required for activation of T_C-cells, and IFN-γ, which induces MHC expression, increases APC activity, activates large granular lymphocytes and together with TNFβ (a lymphotoxin), activates macrophages. Lymphokines (IL-4, 5 and 6) are also required for B-cell activation, leading to the production of anti-graft antibodies. These antibodies fix complement and cause damage to the vascular endothelium, resulting in haemorrhage, platelet aggregation within the vessels, graft thrombosis, lytic damage to cells of the transplant, and the release of the pro-inflammatory complement components C3a and C5a.

The cytokines such as TNF released by T_h-cells, CTLs, NK-cells and macrophages have been shown to mediate direct cytolytic damage to cells. GVH disease can be lethal, causing damage in particular to the skin and gut. It can be avoided by careful matching of the donor and recipient, removal of all T-cells from the graft and immunosuppression.

10.1.5 Rate of Rejection

The rate of rejection depends in part on the underlying effector mechanisms.

Hyperacute rejection

In **Hyperacute rejection** of transplant, rejection is so quick that the graft tissue never becomes vascularized. These reactions are caused by pre-existing host serum antibodies specific for antigens of the graft. Such reactions occur very rapidly in patients who already have antibodies against a graft. Anti-HLA antibodies are produced by prior blood transfusions, multiple pregnancies, or the rejection of a previous transplant. Antibodies against the ABO blood group system can also cause hyperacute rejection. Because humans have preformed IgM and IgG natural antibodies to animal cells, hyperacute rejection prevents transplantation of animal organs to man.

The antigen-antibody complexes that form activated complement system result in an intense infiltration of neutrophils into the graft tissue. The resulting inflammatory reaction causes massive blood clots within the capillaries, preventing vascularization of the graft.

Hyperacute rejection occurs within minutes of transplantation and it occurs in individuals with pre-existing humoral antibodies either due to blood group incompatibility or free sensitization to Class I MHC. In this type of rejection the transplant never becomes vascularized. The complement system is activated due to antigen antibody complex formation at the graft site, resulting in an intense infiltration by neutrophils.

Acute rejection

Acute rejection is cell-mediated allograft rejection that begins about 10 days after transplantation, takes days or weeks to manifest and is due to the primary activation of T-cells and the consequent triggering of various effector mechanisms. An accelerated cell mediated rejection response is seen if a transplant is given to someone who has been pre-sensitized to antigens on the graft, because a secondary reactivation of T-cells occurs.

There is a massive infiltration of macrophages and lymphocytes at the site of tissue destruction, suggestive of T_h-cell activation and proliferation. There is a rupture of peritubular capillaries seen.

Acute late rejection occurs from 11 days onwards in patients administered with immunosuppressive drugs such as prednisone and azathioprine. This is caused by deposition of immunoglobulins in the blood vessel walls, which induces platelet aggregation in the glomerular capillaries leading to renal shutdown. There is also destruction of antibody-coated cells through antibody-dependent cell mediated cytotoxicity.

Chronic rejection

Chronic rejection is a slow process and develops months or years after acute rejection reactions have subsided. It depends on the genetic disparity between donor and recipient, and the use of immunosuppressive treatment.

These reactions include both humoral and cell-mediated responses by the recipient. Chronic rejection reactions are often difficult to manage with immunosuppressive drugs and may necessitate another transplantation. It is associated with sub endothelial deposits of immunoglobulin and C3 on the glomerular basement membranes.

To minimize rejection of grafts, matching of MHC is done. Various antigens that determine histocompatibility are encoded by 40 different loci of HLA complex in human beings.

The endothelial cells express major histocompatibility complex and stimulatory molecules on the surfaces, enabling them to activate T-cells and to act as targets of cellular and humoral immunity. Direct recognition of donor MHC molecules expressed on endothelial cells by recipient alloantigen-specific CD8 effector T-cell can result in direct killing of endothelial cells and ultimately lead to graft loss. Antibody binding to endothelium can also initiate activation of the complement cascade and/or facilitate macrophage mediated or NK-cell mediated cytotoxicity. This can result in apoptosis or lysis of the endothelial cells, the loss of vascular integrity with resulting graft loss.

Early chemokines are quickly produced in allografts and isografts with similar kinetics and to an equal degree. The later chemokine cascade appears in allografts that direct the recruitment of alloantigen-primed T-cells into the grafts. The early Chemokine cascade directly influences the appearance and intensity of the latest cascade.

Pro inflammatory cytokines, including TNF-α and IL-1, are produced by graft endothelial and parenchymal cells, which stimulates the vascular endothelium and in certain cases the graft parenchymal cells to produce neutrophil and macrophage attractant chemokines, including IL-8, Gro-α, MIP-2 and MCP-1 at early times after the transplant. TNF-α and IL-1 up regulate adhesion molecules and MHC molecule expression on the vascular endothelium of tissues and is likely to facilitate rejection of grafts from such donors.

10.1.6 Grafts that are Not Rejected

Certain areas of the body, like the anterior chamber of the eye, the cornea, and the brain, lack effective lymphatic drainage. Although antigens derived from grafts put in these **privileged sites** may reach the lymphoid tissues, cytotoxic effector cells cannot reach the graft, and these grafts survive fairly well. Cornea is considered to be immunologically privileged site because of absence of blood vessels and lymphatic vessels. Therefore, corneal grafts survive without the need for immunosuppression. At the site there are locally produced immunosuppressive factors (TGF-β and IL-1RA), limited expression of MHC antigen and the presence of Fas Ligand (FasL). The presence of FasL on corneal epithelial cells contributes significantly to graft survival.

Allogeneic **sperm** can successfully and repeatedly penetrate the female reproductive tract without provoking an immune response. Seminal plasma is immunosuppressive, and sperm exposed to this fluid are nonimmunogenic. Prostatic fluid, which is an immunosuppressive component of the seminal plasma, also inhibits complement.

During **pregnancy**, since the foetus possesses antigens derived from its father, it is an allograft within the mother. Though the uterus is not a privileged site, the foetus successfully establishes and maintains itself through pregnancy, in spite of the great histocompatible differences. The immunological destruction of the foetus is prevented by the combined activities of several immunosuppressive mechanisms. There are no MHC molecules expressed on the preimplantation embryos or oocytes. After placenta formation, the foetus is protected from the immune system by the trophoblast. Cells within the trophoblast make HLA-G, a nonpolymorphic Class IB molecule, which fails to trigger a T-cell response and protects the cells against NK-cell mediated lysis. The trophoblast cells do not express MHC Class II molecules. Secondly, the foetus is a source of locally active immunosuppressive factors including the hormones estradiol and progesterone and possibly also chorionic gonadotropin. A major protein in foetal serum, α-fetoprotein, some pregnancy associated glycoproteins, a trophoblast-derived interferon (IFN-ω), and phospholipids in amniotic fluid have immunosuppressive properties. Thirdly, blocking antibodies may be produced in response to foetal antigens. These coat the placental cells, masking antigens and thus, preventing their destruction by maternal T-cells. The immunosuppression generated by the foetus is very local in nature.

A **tissue grown in culture or stored frozen** has enhanced potential for successful transplantation. A thyroid tissue grown in an organ culture for 25 days before transplantation, survives in a recipient as if it were an isograft. Uncultured thyroids are rejected in about 10 days. During culture, there is a loss of dendritic cells from the thyroid. Pre-treatment of donors with cytotoxic drugs or antiserum against dendritic cells reduces subsequent allograft rejection. Each cell population has a distinct ability to provoke an allograft rejection, and the most immunogenic components are the MHC Class II *passenger* leukocytes.

10.1.7 Immunologically Favoured Organs

There are some immunologically favoured organs. On some occasions, individuals who have received a liver allograft do not mount a response against it, whereas others may mount a weak and easily suppressed response. Liver allografts may, by their presence, protect another allograft, like a kidney, from the same donor resulting in its prolonged survival. For kidney transplant, patients are partially immunosuppressed at the time of transplantation. The combination of azathioprine and prednisone is commonly given in the long-term management of kidney grafts. Cyclosporine A given in place of azathioprine also yields good results.

The survival figures for heart transplants have reached 80 per cent, with the combined treatment of azathioprine and prednisone after the three months of cyclosporin therapy.

Individuals suffering from aplastic anemia and immune deficiency disorders are given bone marrow transplants. Acute leukaemia patients who undergo whole body irradiation and chemotherapy to kill the cancerous cells can also go in for bone marrow transplants. For a successful bone marrow transplant siblings offer the best chances of finding a matched donor.

10.1.8 Prevention of Rejection

The matching of donor and recipient MHC molecules, especially for MHC Class II molecules, can reduce graft rejection responses. Not all parts of the graft need to be attacked for rejection to occur. The critical targets are the vascular endothelium of the microvasculature and the specialized and parenchymal cells of the organ, such as renal tubules, pancreatic islets of Langerhans or cardiac myocytes.

Tissue matching can reduce the rejection response. The perfectly matched donor and recipient would be isogeneic, for example monozygotic twins. However this situation is very rare and only the major antigens (MHC, i.e., HLA) can be practicably matched. Nowadays, sensitive and accurate typing is achieved using the polymerase chain reaction (PCR) to identify HLA genes in the DNA of donors and recipients. A good organ graft survival is obtained when the donor and recipient share the same MHC Class II antigens, especially HLA-DR because these are the antigens that directly activate the recipient's T_h-cells.

10.2 IMMUNOSUPPRESSION

The techniques for inhibiting an immune response (known as **immunosuppression**) may be classified into three groups. The most widely used technique involves drugs or radiation that inhibit cell division, and therefore, reduce the multiplication of antigen-sensitive cells upon encountering an antigen. A second technique selectively eliminates T-cells by means of anti-T-cell serum, by monoclonal antibodies, by selective drugs, or by thoracic duct drainage. A third way can be a manipulation of the natural control mechanisms of the immune responses. There are two main categories of immunosuppressive treatment: antigen non-specific and antigen specific immunosuppression.

10.2.1 Non-specific Immunosuppression

The non-specific immunosuppression blunts or abolishes the activity of the immune system regardless of the antigen. This can leave a graft recipient very vulnerable to infections. Radiation, corticosteroids, cytotoxic drugs and cyclosporin are the non-specific immunosuppressive agents that are routinely used.

Radiation

Lymphocytes are extremely sensitive to X-rays. X-rays can therefore be used to eliminate lymphocytes in the transplant recipient just before grafting. The typical protocol is daily x-irradiation treatments of about 200 rads per day for several weeks until a total of 3400 rads has been administered. The recipient is grafted in this immunosuppressed state. Because the bone marrow is not x-irradiated, lymphoid stem cells proliferate and renew the population of recirculating lymphocytes. The newly formed lymphocytes appear to be more tolerant to the antigens of the graft.

X-irradiation exerts its effects on cells by several mechanisms. First is a simple technique through ionizing rays hitting an essential unique molecule, like the DNA within the cell. Another mechanism involves the formation of reactive free oxygen and hydroxyl radicals as a result of ionization in aqueous solutions. These free radicals can react with dissolved oxygen to form peroxides that have toxic effects on many cell processes.

Corticosteroids

Steroid receptors are found inside the cell and act by regulating DNA transcription. Steroids are absorbed directly into a cell, where they bind to receptors in the cytoplasm. The steroid-receptor complexes are then transported to the nucleus, where they bind to the chromatin protein and the DNA. As a result of derepression of RNA transcription, new mRNA is formed and initiates new protein synthesis. These newly synthesized proteins mediate the effects of corticosteroids. Corticosteroids act in four areas. They effect leukocyte circulation; they influence the immune effector mechanisms in lymphocytes; they modulate the activities of inflammatory mediators; and they modify protein, carbohydrate and fat metabolism. In humans, the number of circulating eosinophils, basophils, and lymphocytes decline abruptly when steroids are administered. Corticosteroids suppress the cytotoxic and phagocytic abilities of neutrophils.

Steroids have anti-inflammatory properties and suppress activated macrophages, interfere with the function of Antigen-Presenting Cells (APCs) and reduce the expression of MHC antigens. Effectively steroids reverse many of the actions of IFN-γ on macrophages and transplanted tissues.

Corticosteroids, like prednisone and dexamethasone are potent anti-inflammatory agents that show their effects at many levels of the immune response.

Cytotoxic drugs

The majority of immunosuppressive drugs have been designed to inhibit cell division. They act on various stages of nucleic acid synthesis and activity. Cyclophosphamide, azathioprine, and methotrexate are highly toxic and fairly non-specific in their actions. These days, two very specific and relatively non-toxic drugs, cyclosporin and FK506 are being used in the treatment of allograft rejection. Alkylating agents, like cyclophosphamide, cross-link DNA helices and prevent their separation, and thus, inhibit cell division and gene transcription. However, they are toxic for both resting and dividing cells, especially for dividing immunocompetent cells. Unlike cyclophosphamide, azathioprine affects only proliferating lymphocytes and not resting cells. It is therefore a less potent immunosuppressant and can suppress both primary and secondary antibody responses only if given after antigen exposure. FK506 and rapamycin too have immunosuppressive properties. FK506 suppresses lymphokine production by T_h-cells, while rapamycin interferes with the intracellular signalling pathways of the IL-2 receptor, preventing IL-2 dependent lymphocyte activation. Azathioprine is an antiproliferative drug, an analogue of 6-mercaptopurine. It incorporates into the DNA of dividing cells preventing further proliferation. It is often given just before and after transplantation to diminish T-cell proliferation in response to the alloantigens of the graft. Methotrexate is the folic acid antagonist and blocks purine biosynthesis. Other drugs used are hydroxyurea, which blocks DNA synthesis; and 5-fluorouracil, which is a pyrimidine analogue.

Cyclosporin

Cyclosporin is a selective immunosuppressive polypeptide that blocks the helper T-cell response without affecting resting lymphocytes. It inhibits the production of IFN-γ by helper T-cells and hence, prevents IFN-γ induction of MHC Class I expression in grafts. Cyclosporin suppresses lymphokine production by T_h-cells and reduces the expression of receptors for IL-2 on lymphocytes undergoing activation. Cyclosporin A and FK506 blocks activation of resting T-cells by inhibiting the transcription of genes encoding IL-2 and the high affinity IL-2 receptor, which are essential for activation. The combination of corticosteroids and cyclosporin is especially potent and can give essentially 100 per cent survival of renal grafts.

The non-specific immunosuppressive agents can be used to block transplant rejection, but these may also reduce resistance to infections. Currently, three non-specific agents that are widely used are steroids, cyclosporin and azathioprine.

10.2.2 Specific Immunosuppression

Specific immunosuppression reduces anti-graft responses without increasing susceptibility to infection. It is used to inactivate only those lymphocyte clones, which cause graft rejection. Such immunosuppression is caused by Anti-Lymphocytic Serum (ALS) that suppresses the cell-mediated immune response and leaves the humoral immune response relatively intact.

Monoclonal antibodies against lymphocyte surface molecules, especially CD3, CD4, CD8, and the IL-2 receptor, can be used to eliminate cells or to block their function. Cytotoxic drugs are attached to these antibodies to increase their effectiveness.

10.2.3 Antibodies as Immunosuppressive Agents

Depending upon their specificity, some antibodies can block the formation of other antibodies in various ways. They can do so by (i) hindering the reaction of the immunogen with appropriate cell surface receptors, and (ii) inactivating lymphocytes especially T-cells.

Suppression by Antibody as Anti-immunogen

If antibodies are added to the immunogen before its injection (or are present in the animal due to previous immunization), the induction of antibody formation is blocked. Inhibition increases when antibody is in excess. Haemolytic disease of the newborn, which is due to maternal antibodies against foetal Rh red cells, can be reduced in incidence by injecting anti-Rh antibodies into Rh^- mother at the time of delivery of an Rh^+ baby. There is no anti-Rh anamnestic response in a subsequent pregnancy with an Rh^- baby as the mother was prevented from being immunized. In case of measles, maternal antibodies against measles can cross the placenta. Hence, immunization of the newborn is deferred till six to seven months of age.

Suppression by Antibody as Anti-receptor

Under some circumstances anti-immunoglobulins suppress the production of immunoglobulins either by (i) eliminating B-cells with the target immunoglobulin as surface receptor or (ii) blocking their stimulation by antigen. Depending upon the specificity and conditions of administration, the anti-Ig suppresses the formation of immunoglobulin of the corresponding class (isotype, allotype or idiotype).

Suppression by Anti-Lymphocyte Serum (ALS)

ALS is usually prepared in horse or rabbit against another species thymus cells, white blood cells or cultured lymphocytes. The Anti-Lymphocyte Globulin (ALG) causes preferential inactivation of the long-lived pool of the recirculating lymphocytes (which are mainly T-cells). This prolongs survival of allografts. It also reduces antibody formation in primary responses to many independent antigens, if given just before the antigen.

Monoclonal Antibodies

Monoclonal antibodies directed against various surface molecules on cells of the immune system are used to suppress T-cell activity in general or to suppress the activity of subpopulations of T-cells. This involves the use of monoclonal antibody to the CD3 molecule. Injection of such monoclonal antibodies results in a rapid depletion of mature T-cells from the circulation. Monoclonal antibody therapy is also used to treat donor's bone marrow before it is transplanted. This is done to deplete the immunocompetent T-cells in the bone marrow transplant, which react with the recipient tissues, causing **graft-versus-host disease**.

Some newer immunosuppressive drugs are under investigation and are yet to undergo clinical trials. One is FTY720, which has a dramatic effect on lymphocytes migration, leading to marked sequestration of lymphocytes in the lymphoid tissues and a sharp fall in the circulating numbers of these cells.

MULTIPLE-CHOICE QUESTIONS

1. Transplanted cells are mainly destroyed by:
   (a) neutrophils (b) macrophages
   (c) B-cells (d) T-cells
2. The foetus can be considered:
   (a) an isograft (b) an allograft
   (c) a heterograft (d) antigraft
3. Corneal grafts are not rejected because they:
   (a) do not possess histocompatibility antigens
   (b) are resistant to lymphocytotoxic activity
   (c) have no lymphatic drainage
   (d) are not exposed to antibodies
4. One of the ways by which allograft rejection is prevented is through the administration of:
   (a) antibodies to CD3 (b) immunoglobulin E
   (c) anti-interferon antibodies (d) anti-rhesus antibodies
5. Corticosteroids mainly suppress allograft rejection by suppressing:
   (a) macrophage function (b) T-cell function
   (c) antibodies synthesis (d) neutrophil function
6. The major targets of cytotoxic T-cells within a kidney allograft are:
   (a) neutrophils (b) proximal tubule cells
   (c) vascular endothelial cells (d) macrophages
7. The major clinical problem associated with bone marrow allografts in humans is:
   (a) aplastic anaemia (b) allograft rejection
   (c) contact dermatitis (d) graft vs host disease

REVIEW QUESTIONS

10.1 Why is the foetus not rejected by its mother?

10.2 Explain the act of the specific immunosuppressive therapies used during transplantation.

10.3 Write notes on:
   (i) Graft vs host reaction
   (ii) Rate of rejection of grafts
   (iii) Immunologically favoured organs

11

TUMOUR IMMUNOLOGY

11.1 TUMOUR

A cell can sometimes become altered in such a way that it grows uncontrollably and gives rise to a tumour. A tumour is an abnormal growth of cells that form a lump or mass of transformed cells, which occurs due to loss of regulatory control on cell division; the genes responsible for certain characteristics of the cell undergo mutation. Some of these cancerous cells may be sufficiently abnormal, so that they can trigger an immune response. At times, such an immune response can destroy the tumour.

The immune system recognizes these changed surface molecules and attacks the tumour cells as though they are foreign invaders. However, tumour cells still survive and thrive in the body by evading the immune response. Evasion by these cancerous cells is achieved by changing the antigenic molecules repeatedly and thereby confusing the immune system.

11.1.1 Causes of Tumours

As a result of exposure to certain chemicals, virus infection, or mutation, a cell may not have normal regulation of cell division. At the same time it may develop the ability to invade places that are normally occupied by other cells. A cell that is proliferating in an uncontrolled fashion gives rise to a growing clone of cells that eventually develops into a tumour, or **neoplasm.** There are two types of tumours: **benign tumour** and **malignant tumour**. Benign tumour is usually encapsulated and confined to a certain site, whereas in a malignant tumour the cells break off from the main tumour mass and travel through the blood system lodging themselves elsewhere and initiating new tumour formation at the site where they have lodged. Cells of the malignant tumour expressed special receptors on their membranes for autocrine factors, which endow the cancer cells with special antigenic properties.

The secondary tumours that arise in these distant sites are called **metastases.** Tumours arising from epithelial cells are called **carcinomas,** and those arising from mesenchymal cells such as muscle, lymphoid, or connected tissue cells are called **sarcomas. Leukemia** is the tumour derived from haemopoietic cells.

The essential difference between a normal cell and tumour cell is loss of control of cell division or of apoptosis. Tumour cells may also show other abnormalities, for instance, the proteins on the surface of a tumour cell may be different from those found on normal cells. Malignant tumours behave abnormally due to expression of mutated or viral genes and/or deregulated expression of normal genes of a cell. Cancer cells express proteins that either are not expressed at all or are present in much lower quantities in normal cells. The tumour cells express antigens that can stimulate immune response in the host. These new or altered proteins may be recognized by the cells of the immune system and therefore, trigger off an immunological attack.

A normal cell develops into a cancerous cell usually through a multistep process of clonal evolution that is driven by a series of somatic mutations. There is a progressive conversion of the normal cell to a precancerous cell and finally into a cancerous cell.

In an experiment, a sarcoma induced by methylcholanthrene was excised and transplanted into syngeneic mice. The surgically cured host was immunized with X-irradiated cells of its original tumour (the X-irradiation damaged chromosomes, but left the cell antigens intact). The immunized animal could reject a graft of its own tumour. The resistance could be transferred to normal syngeneic recipients with its T-cells.

The tumour antigens are detected by rejection of transplanted tumour cells, and hence are called **tumour rejection antigens** or **tumour transplantation antigens.** A remarkable feature of these tumour rejection antigens is an enormous diversity. A sarcoma induced by MethylCholAnthrene (MCA) will not induce protective immunity against another MCA induced sarcoma, even if both the tumours are obtained from the same mouse. Therefore, they are called Tumour Specific Transplantation Antigens (TSTAs).

If tumour cells are sufficiently different from normal, they'll be regarded as foreign and stimulate an immune response. The major mechanisms of tumour cell destruction involve NK-cells, cytotoxic T-cells, activated macrophages and antibodies.

Tumour immunology has its base on two simple propositions: (i) Tumour cells have distinctive antigens, which are known as Tumour Associated Antigens (TAAs). These are present to a negligible extent on normal cells, including those of the same histologic type as the tumour. (ii) Immune responses of the host to these antigens can destroy a tumour cell as though it were a virus-infected cell or an allograft.

Hence, there is an immune surveillance by lymphocytes, especially T-cells, as they circulate throughout the body. They detect and eliminate tumours as they arise. Tumours that succeed in growing and threaten the host's survival might be a result of (i) defects in the host's immune system, (ii) selective growth of tumour cell variants whose TAAs have little or no immunogenicity, or (iii) other mechanisms that enable them to escape tumour destruction.

11.2 TUMOUR ANTIGENS

Two types of tumour antigens have been identified on tumour cells: **Tumour-Specific Transplantation Antigens (TSTAs)** and **Tumour Associated Transplantation Antigens (TATAs).**

11.2.1 Tumour-Specific Transplantation Antigens (TSTAs)

Tumour-specific antigens are unique to tumour cells and do not occur on normal cells in the body. Tumour associated antigens that elicit transplantation rejection responses are called Tumour Specific Transplantation Antigens (TSTAs). Different tumours arising due to a given carcinogen can differ in their TSTAs. Tumours induced by an oncogenic virus share TSTAs.

11.2.2 Tumour Associated Transplantation Antigens (TATAs)

Tumour-associated antigens are not unique to the tumour cell. They may be proteins that are expressed on normal cells during foetal development when the immune system is immature and unable to respond; and these are not normally expressed in the adult. In tumour cells there is a reactivation of the embryonic genes. Tumour-associated antigens may also be proteins that are normally expressed in very low levels on normal cells, but are expressed in much higher levels on tumour cells.

Tumour Associated Transplantation Antigens (TATAs) are of two types. The first is an antigen, which is shared by many tumours, even though it may not be even of the same tissue of origin. The second is an antigen, which is specific to the individual tumour and is called tumour specific transplantation antigens. Tumours may express both specific and shared antigens.

11.3 TUMOUR ANTIGENS RECOGNIZED BY T-LYMPHOCYTES

These tumour antigens may stimulate T-cell mediated rejection of tumour transplants in animals previously immunized with the tumour. These antigens are cell proteins that MHC complexes to either $CD4^+$ or $CD8^+$ T-lymphocytes. Some tumour antigens recognized by T-lymphocytes are unique and some are present on many cancers of different types involving different tissues.

Tumour antigens that are recognized by human T-cells fall into one of the four following major categories. They may be antigens that are:

1. Encoded by genes specifically expressed by tumours.
2. Encoded by variant forms of normal genes that have been altered by mutation.
3. Normally expressed only at certain stages of differentiation or only by differentiation lineages.
4. Over expressed in particular tumours.

11.3.1 Tumour Antigens Encoded by Normally Silent Cellular Genes in Spontaneous Tumours (Tumour Specific Shared Antigens (TSSAs))

Some genes are usually silent and do not express in normal cells or they are expressed in the early stage of development, during differentiation and embryogenesis, before the onset of self-tolerance mechanism. If, as a result of malignant transformation of a cell, these genes are switched on, they are expressed inappropriately in the wrong tissue at the wrong time in the life of an individual and then they behave as tumour antigens and evoke immune responses. These tumour antigens may be shared by many different tumours.

11.3.2 Tumour Antigens Coded by Oncogenes or Tumour Suppressor Genes

This class of tumour antigens includes non-mutated oncogenic molecules whose expression is at least partially responsible for the tumour's malignant phenotype. These proteins are often growth factors and can be expressed at low levels by normal cells. They are constitutively expressed in high levels by malignant cells. Most of the tumours express genes whose products are required for malignant transformations as well as maintenance of the malignant phenotype. These genes have been found to be altered or mutated form of normal cellular genes that regulates cell division and differentiation process. Carcinogen induced point mutations, deletions or chromosomal translocations may change the status of proto-oncogenes to oncogenes whose products have transforming activity. There are other genes called **tumour suppressor genes**, which also code for proteins for normal cellular growth and differentiation. Point mutation at any site on these genes results in introduction of nonfunctional products or it may inactivate the gene with the resulting malignant transformation occurring. The altered products of mutated proto-oncogenes or tumour suppressor gene expressed in tumours may stimulate immune response in the host.

Tumour-specific antigens have been identified on tumours induced with chemical or physical carcinogens and on some virally induced tumours.

Tumours induced by oncogenic viruses tend to gain new antigens, characteristic of the inducing virus. These antigens, although coded for by the viral genome, are not part of the virus particle.

11.3.3 Tumour Antigens Encoded by Genomes of Oncogenic Viruses

RNA and DNA viruses can induce tumour formation in experimental animals as well as human beings. Virally induced tumours usually contain integrated proviral genomes in their cellular genomes and these tumours often express viral genome coded proteins. DNA viruses, such as papova viruses and adenoviruses, induce malignant tumours in neonatal and immunodeficient adult rats. In humans, DNA viruses, such as Epstein-Barr virus (EBV), are associated with B-cell lymphomas, Hodgkins lymphoma and nasopharangeal carcinoma. Human Papilloma Virus (HPV) is associated with most human cervical carcinomas. RNA tumour virus' retroviruses induce tumour with in days after infection and are called acute transforming retroviruses. The examples are Rous sarcoma virus (carrying Src oncogene) and avian myelocytomatosis virus (carrying myc oncogene). The transforming retroviruses bring about mis-regulation of cellular genes by inserting their genome in the cellular genome and bring about loss of cell growth control and differentiation.

11.3.4 Virally Induced Tumours

Virally induced tumours express tumour antigens shared by all tumours induced by the same virus. There is great potential value of these virus-induced tumour antigens. Mice immunized with a preparation of genetically engineered polyoma virus tumour antigen become immune to subsequent injections of live polyoma-induced tumour cells. In another experiment mice were immunized with a vaccinia virus vaccine engineered with the gene encoding the polyoma tumour antigen. These mice too developed immunity against live polyoma-induced tumour cells.

11.3.5 Tumour Antigens Resulting from Mutations

Antigens in this category are peptides derived from regions of proteins that are mutated in tumour cells. Two of these mutations affecting the human genes Cyclin-Dependent Kinase (CDK4) and β-catenin respectively may be involved in oncogenisis. The CDK4 mutation prevents the protein from binding to its inhibitor, p16. This alters the regulation of the cell cycle leading to uncontrolled growth of tumour cells. Another mutation that antagonizes apoptosis has been identified with CTLs specific for a human squamous cell carcinoma. The antigen is encoded by the mutated form of the CASP-8 gene which encodes for the protease caspase-8.

11.3.6 Antigens of Tumours Induced by Chemical Carcinogens

Chemically induced tumours differ from the virus-induced variety by developing new antigens unique to the tumour and not to the inducing chemical (Figure 11.1). Tumours induced by a single chemical in different animals of the same species may be antigenically different; even within a single chemically induced tumour mass, antigenically distinct subpopulations of cells exist. As a result, resistance to one chemically induced tumour does not prevent growth of the second tumour induced by the same chemical.

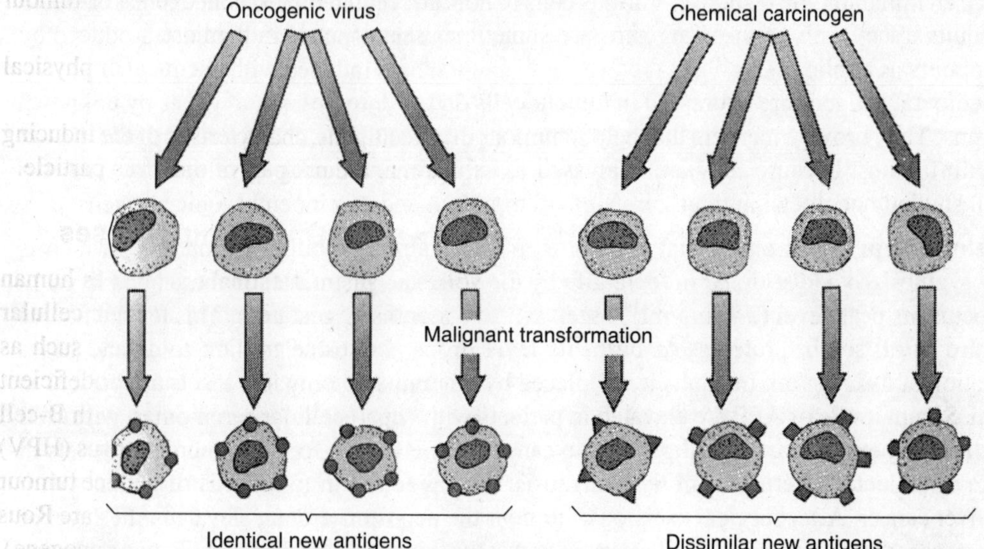

Figure 11.1 The development of new antigens on tumour cells induced by viruses or chemical carcinogens.

Many tumours induced by chemical carcinogens have individually distinctive TAAs that are known as **private TAAs**.

Methylcholanthrene painted on different skin areas of the same host cause multiple fibrosarcomas with little or no cross reactivity. The gene that encodes one of the private TAAs (i) lacks a leader sequence implying that it is not a membrane protein, and (ii) the difference between it and the corresponding gene in other cells is a single nucleotide replacement that changes arginine to histidine. Thus, private TAAs may be peptides that derive from any mutated gene, whether or not the gene product is membrane associated. Tumours can differ widely in immunogenicity and latent period. There are other TAAs that are shared by many tumours, such TAAs are called **public TAAs**. They sometimes represent anomalous expression of normally silent genes, for example Forssman antigen of human gastrointestinal cancer.

11.3.7 Tumour Antigens Defined by Xenogenic Antibodies

There are many tumour antigens present as surface molecules on the tumour cells. These antigens are found in different types of tumours arising from the same type of cells and most may also be found on some normal cells. These antigens do not stimulate immunological response in the host and are called Tumour-Associated Antigens (TAAs). Several classes of these antigens may be expressed on the same tumour. These TAAs are important for diagnosis and possible treatment of cancers. There are three classes of TAAs namely, oncofoetal antigens, altered glycoproteins and glycolipid antigens and tissue specific antigens on tumour cells.

Oncofoetal antigens

These proteins are normally expressed on developing foetal tissues, but not on adult tissues. These antigens appear early in embryonic development, before the immune system acquires immunocompetence. If these antigens appear later on cancer cells, they are recognized as non-self and induce an immunologic response. Various cells in normal foetal development express these. In normal adults trace levels of these antigens are sometimes seen. Some of them are produced by certain tumours as public TAAs.

Oncofoetal antigens are expressed in tumour cells due to derepression of genes by unknown mechanisms. They provide markers that help in tumour diagnosis. The oncofoetal antigens are not antigenic in the host because they are expressed as self-proteins during development. The two most well studied oncofoetal antigens are alpha-fetoprotein and carcinoembryonic antigens.

Alpha-fetoprotein: This oncofoetal antigen is a 70 kD alpha globulin glycoprotein which is normally synthesized and secreted in foetal life by the yolk sac, gastrointestinal tract and liver. In human foetus, its peak level (2–3 mg/ml) is seen at 3 to 6 months of gestation. This level amounts to one-third of all serum proteins. At birth, its level drops about one million folds and has a concentration of 5–25 ng/ml. In adults it is replaced by albumin and only low levels are present in the serum. Serum levels of AFP are elevated in patients with hepatocellular carcinomas, germ cell tumours, and sometimes gastric and pancreatic cancers. AFP is also found in increased levels in the serum of patients suffering from liver cirrhosis. AFP is found in two-thirds of patients with primary liver cancer. After surgical excision of tumour the persistence of an elevated AFP level or a later increase indicates that tumour cells remain or that the tumour has recurred. AFP may represent the foetal form of serum albumin. It is a single polypeptide of molecular weight 70,000 and some amino acid sequence similarity with albumin. It also binds estrogen and a variety of anionic dyes.

Carcinoembryonic antigen: This is a member of the immunoglobulin superfamily and is highly glycosylated. It is 180 kD integral membrane protein. Carcinoembryonic antigen (CEA) is present in trace levels (< 5 μg/ml) in normal human serum. Increased levels of CEA are seen in the serum of most patients with cancer of lower gastrointestinal tract and pancreas. It is seen less frequently in patients with cancer of other organs like breast or lung, or with certain chronic or recurrent inflammatory conditions like colitis and chronic bronchitis. CEA expression is greatly enhanced in colon cancer and it is also released into extracellular fluid. CEA levels are used to monitor the occurrence or recurrence of metastatic colon carcinoma after primary treatment.

Altered glycoprotein and glycolipid antigens

Abnormal surface glycoproteins or glycolipids are expressed on most humans and experimental tumours as a result of defects in the sequential addition of carbohydrate qualities to core proteins or lipid molecules. Abnormal glycolipid and glycoprotein synthesis is associated with tissue invasion and metastatic behaviour of many tumours. This class of TAAs continues to be a preferred target for antibody-based approaches to cancer therapy.

11.3.8 Monoclonal Antibodies to TAAs

Monoclonal antibodies to human TAAs are produced when mice are immunized against cells from human tumours. Supernatant fluids from B-cell hybridomas prepared from these mice are screened for their ability to bind to a wide variety of cells, mainly (i) the tumour cells used as the immunogen, (ii) the normal cells from the individual furnishing these tumour cells, and (iii) large panels of other tumour cells of different types from other individuals. However, these antibodies cross-react with antigens of some normal cells to a very limited extent that is almost negligible. Monoclonal antibodies can be used to **locate and identify** tumour masses in case of metastasis of primary tumours, and to aid in the **destruction** of tumours.

11.4 HOST IMMUNE RESPONSES TO TUMOURS

In a tumour-bearing host, TAAs elicit diverse immune responses, involving all classes of antibodies and types of T-cells. The tumour cells can be destroyed by

(i) Antibodies, with or without the aid of complement. This is antibody-dependent cell-mediated cytotoxicity.
(ii) CTLs or cytotoxic T-lymphocytes.
(iii) Macrophages activated by lymphokines.
(iv) Natural killer cells.

For unknown reasons, tumour cells are more susceptible than normal cells to destruction by all of these aforementioned effector mechanisms.

$CD4^+$ and $CD8^+$ T-cells can be readily grown in vitro from surgically removed human tumours that are cultured in presence of IL-2. The cells are highly active, can lyse ^{51}Cr-labelled cells from the tumour and are called **Tumour Infiltrating T-lymphocytes (TIL)**. These tumour-infiltrating T-lymphocytes can reduce the mass of some tumours substantially through a specific immune attack on the tumour cells.

11.4.1 Immune Response to Tumour Antigens

Both humoral and cell mediated immune responses to tumour antigens have been seen in vivo. The main effector mechanisms in anti-tumour immunity is due to T-lymphocytes, natural killer cells, macrophages and antibodies.

Cytotoxic T-Lymphocytes (CTLs) provide effective anti-tumour immunity in vivo. CD8 CTLs mediated rejection of transplanted tumour is an example of specific anti-tumour immunity in vivo. CTLs may perform immune surveillance by recognizing and killing potentially malignant cells that express peptides coded by mutant cellular genes and are present in association with Class I

MHC molecules. CTLs can be isolated from animals and humans who already possess well established tumours. Mononuclear cells found in human tumours as infiltrates are called Tumour Infiltrating Lymphocytes (TILs) discussed above, also include CTLs with a capacity to lyse the tumours from which they were derived.

CD4 helper T-cells have an important role in the anti-tumour response by enhancing the secretion of cytokines for effective development of CTLs. CD4 cells also secrete tumour necrosis factor and interferon gamma on activation by tumour antigens which can increase tumour cell Class I MHC expression and make the cells sensitive to lysis by CTLs.

Natural Killer cells (NK-cells)

These are produced as natural acquired immune response to tumours. The mechanism of lysis of tumour cells by NK-cells is the same as that of CTLs. The NK cells do not express T-cell antigen receptors, and kill targets in an MHC unrestricted manner. The activity of NK-cells is enhanced by cytokines including interferons, TNFα, IL-2 and IL-12. NK-cell activity depends on the concurrent stimulation of T-cells and macrophages. Lymphokine Activated Killer (LAK) cells are derived from peripheral blood cells or TILs from tumour patients with high doses of IL-2. LAK cells lyse other cells, including tumor cells, in a nonspecific manner.

Macrophages

Macrophages preferentially lyse the tumour cells and possess receptors which can be targeted to tumour cells coated with antibodies. They kill tumour cells by various mechanisms, which include the release of lysosomal enzymes, reactive oxygen metabolites and nitric oxide. Activated macrophages also secrete the cytokine tumour necrosis factor which is capable of selectively killing tumour cells and not normal cells. TNFα can kill by production of free radicals or by disrupting the cytoskeletal proteins and gap junction proteins.

Antibodies

Tumour bearing hosts produce antibodies against tumour antigens even though T-cells mediate effector anti-tumour immune responses. In some cases the antibody production is specific for viral antigens. Patients with Epstein-Barr virus associated lymphomas show antibodies against EBV coded antigens expressed on the surface of their tumour cells. Cancer patients produce antibodies against their own tumours and these antigens are also present on certain normal tissues as well. However, there is no evidence for a protective role of such humoral antibody responses against tumour development and growth.

11.5 IMMUNE SURVEILLANCE

In 1950's Paul Ehrlich coined the term *immune surveillance* for the immune system, which is continuously looking for abnormal cells and their antigens in the host and destroy these abnormal cells before they grow into tumours or could kill tumours after they are formed. Immune response therefore, is important in preventing the spread of potentially oncogenic viruses rather than surveillance against all tumours. It is likely that immune responses to tumours are often weak and relatively late.

Individuals with growing tumours often produce antibodies and T-cells that react with cells and cell-free components of the tumour. They either destroy tumour cells or interfere with their growth and spread. This implies that people who are chronically immunodeficient have higher incidence of cancer. Evidence argues that the normal immune responses have the ability to reduce the frequency of certain rare tumours, but they cannot control the more frequent forms of human cancer.

11.6 ESCAPE FROM SURVEILLANCE

Although an individual has innate ability to build an immune response against cancer, death due to cancers is because of failed immune response. Tumour cells have the ability to evade the immune system. Why or how the tumours escape immune surveillance is not fully understood. However, tumours have multiple mechanisms for evading immune responses. The most obvious is that the tumour may not be immunogenic. This may be not because potential tumour antigens are lacking, but because the tumour cells are poor Antigen Presenting Cells (APCs). Tumour cells may also lack molecules required for adhesion of lymphocytes such as LFA-1 and -3, or they may express molecules, such as mucins; which can be anti-adhesive. They may also secrete immunosuppressive cytokines such as TGF-β. An important escape mechanism is loss of MHC antigens, leading to inability to present tumour antigen peptides. More than 50 per cent tumours may lose one or more MHC Class I alleles, and sometimes all Class I is lost. Other tumours are covered by large amounts of a polysaccharide or a glycoprotein. Moreover, the immune surveillance acts on a heterogeneous cell population and eliminates the tumour with the most immunogenic TAAs, leaving those with weak TAAs to proliferate.

Anti-tumour antibodies themselves act as blocking factors in some tumours. Probably the antibody binds to tumour specific antigens and mask the antigens from cytotoxic cell action. Certain tumour specific antigens disappear from the tumour cells in the presence of serum antibody and then reappear after the antibody is no longer present. This phenomenon is called **antigenic modulation**. A co-stimulatory signal is triggered by the interaction of B7 molecule of APCs with CD28 on T-cells together with interaction of TCR with antigen-MHC complex. Both the signals induce the T-cells to produce IL-2 which also helps in proliferation of T-lymphocytes. In most of the individuals suffering from tumours, poor immunogenicity to tumours may be largely due to lack of co-stimulatory signals. T-cell anergy may also occur in individuals suffering from cancer due to lack of sufficient number of APC's in the vicinity of tumour and improper T-cell activation.

Cancers elicit protective immunity in the primary host. Tumour immunity depends on many factors. The degree of tumour immunity depends upon the type of cancer and method of its induction, or lack of induction. UV-induced cancers are highly immunogenic, methylcholanthrene-induced tumours less so, and spontaneous tumours even less so. The primary animal develops immunity to subsequent immune challenge. Protective immunity is only seen in prophylactic immunization—once a mouse has been implanted with tumour, immunization with irradiated cancer cells derived from the growing tumour does nothing to mitigate tumour growth. It is also not possible to test immunogenicity of tumours in humans.

11.7 THERAPEUTIC STRATEGIES

There are many general approaches used in attempting to cure or slow tumour growth. Active immunization against tumours can be done by injecting the killed or irradiated tumour cells together with non-specific adjuvants. This induces active immunization against the particular tumour. Immunization can also be done with protein or with DNA encoding tumour antigen/peptides as an immune therapy of cancer. The DNA can be introduced either through direct injection into the target tissue or following immunization with Dendritic Cells (DC) that are tranfected with tumour antigen genes into DC. The advantage of having DNA immunization is that DNA can make entry into a cell and can mimic natural infection.

In adoptive cellular therapy, cultured immune cells that have anti-tumour reactivity are introduced into the tumour bearing host. Lymphokine activated killer cells, tumour infiltrating lymphocytes and enhancement of APC activity for immune therapy of cancer are used.

Some other therapeutic approaches against cancer include the following.

Surgery

It is often done to remove the primary tumour mass. This is followed by **X-irradiation and chemotherapy** to destroy the cells that might have spread.

Monoclonal antibodies to TAAs

Immunotoxins are made by covalently attaching toxic polypeptides to monoclonal antibodies. These immunotoxins are targeted to specific tumours to kill them. When using passive therapy with anti-tumour antibodies, the approach is to selectively kill tumour cells. Anti-tumour antibodies coupled with radioisotopes and drugs have been tried for immune therapy of cancer in humans. For the treatment of B-cell lymphomas that express immunoglobulins of a particular idiotype on the surface, anti-idiotypic antibodies and monoclonal antibodies are used.

Lymphokines-activated killer cells

CTLs activated by IL-2 are injected intravenously to patients to reduce tumour mass. Killer cells are obtained by incubating peripheral blood lymphocytes with IL-2 at increased concentration. Activity of these Lymphokines Activated Killer cells (LAK cells) is like that of NK-cells. However, this has life-threatening side effects.

Tumour infiltrating lymphocytes

Tumour infiltrating lymphocytes are $CD4^+$ and $CD8^+$ T-cells. They reduce the mass of some tumours substantially through a specific immune attachment on the tumour cells. Monoclonal antibodies can also be used to prepare tumour specific immunotoxins. Here, monoclonal antibodies against tumour associated antigens are linked to a toxin of bacterial origin.

Targeting tumour cells for destruction by cytotoxic lymphocytes

Heteroduplex antibody conjugates are formed for targeting tumour cells. For example mAb to the invariant CD3 moiety of the antigen specific T-cell receptor (CD3/TcR) is covalently linked to the monoclonal antibody to a TAA (X). Anti-CD3-anti-X cross-links mature T-cells having surface antigen X. The cross-linked T-cells that are CTLs will be activated via aggregation of their CD3/TcR complexes. In this way the CTLs are targeted to kill X-bearing tumour cells.

Heteroconjugate antibodies are made by covalently coupling an antibody specific for tumour antigen to an antibody directed against the surface protein of a cytotoxic effector cell, such as NK-cells or CTLs. These are then targeted against tumour cells. This promotes binding of these effector cells to tumour cells. A hetero to conjugate consisting of anti-CD3 antibody coupled to antibody against a cell surface protein has been used to enhance CTL mediated lysis of the target cell. The anti-CD3 antibody serves to bring the CTL into contact with the target cell and also activate the CTL.

In vitro depletion of bone marrow tumour cells is done by antibody plus complements mediated lysis.

Immune therapy for human cancers. Animal models are limited in the translation of therapy. Antibodies have been successfully used and vaccination can also be used to treat cancer. The idiotypes of B-lymphomas have been used as vaccines. Immunization with irradiated intact tumour cells has been tested against many cancers. Tumour lysate immunization is still being tested. Defined MHC I epitope immune therapy does not lead to clinical benefit. Adoptive immune therapy using T-cells is being done. Immune therapy with peptides pulsed dendritic cells has shown early promise. Inhibition of down regulation or immune modulation may be clinically valuable. The most successful treatment approaches are however based on individually distinct tumour antigens. Some cytokines like the interferous (α, β and γ), IL-1, IL-2, IL-4, IL-5, IL-12, GM-CSF and TNFα have been evaluated in cancer immunotherapy. But systemic administration of cytokines has serious side effects, and thus these are not approved for general practice.

MULTIPLE-CHOICE QUESTIONS

1. One mechanism by which tumours evade immunological destruction is:
   (a) release of lymphotoxins
   (b) production of immunosuppressive molecules
   (c) altered pathway cytotoxicity
   (d) secretion of anticomplementary factors
2. Carcinoembryonic antigen is characteristically secreted by tumours of the:
   (a) kidney (b) lungs
   (c) bones (d) gastrointestinal tract
3. Macrophage anti-tumour activity is mainly mediated by:
   (a) tumour necrosis factor and interleukin 1
   (b) nitric oxide and interleukin 6
   (c) nitric oxide and tumour necrosis factor
   (d) interleukin 1 and interleukin 6
4. Tumour enhancement is:
   (a) promotion of tumour growth by antibody
   (b) promotion of tumour growth by drugs
   (c) promotion of tumour growth by NK-cells
   (d) inhibition of tumour growth by antibodies

REVIEW QUESTIONS

11.1 What is the biological significance of NK-cells?

11.2 Write notes on:

(i) Tumour associated antigens
(ii) Oncofoetal antigens
(iii) Immune surveillance
(iv) Escape from immune surveillance
(v) Monoclonal antibodies in cancer treatment

11.3 What are the therapeutic strategies that are used to combat tumours?

12

AUTOIMMUNITY

12.1 AUTOIMMUNITY

When the immune system works against the cell components of the body it is called autoimmunity. Since the origin, the T- and B-lymphocytes are conditioned to discriminate between the self and non-self. When the cells fail to discriminate between self and non-self components it leads to autoimmune diseases. During the process of selection those lymphocytes which show reaction to self components are committed to clonal anergy or apoptosis, and those committed to foreign antigens are selected to go through clonal expansion.

It was earlier believed that normal, healthy animals were unable to mount any immune response against self-antigens because of self-tolerance. However, animals can produce autoantibodies easily when their immune systems are appropriately stimulated. This is because not all self-reactive lymphocytes are deleted during T-cell and B-cell maturation. Normal healthy individuals possess mature, recirculating, self-reactive lymphocytes. A small number of B-cells, reactive to normal tissue antigens, are always present in lymphoid organs. Presence of the self-reactive lymphocytes, however, does not inevitably result in autoimmune reactions. Their activity must therefore, be regulated in normal individuals through clonal anergy or clonal suppression. A breakdown in this regulation can lead to activation of self-reactive clones of T- or B-cells thus, generating humoral or cell-mediated responses against self-antigens. Stimulation of these cells provoke the appearance of autoantibodies in serum, which are directed against common autoantigens, for example DNA, IgG, phospholipids, red blood cells, and lymphocytes. These antibodies can react with normal tissues, but usually have no adverse effects. Sometimes the regulation of these auto reactive cells may break down. When this happens, clones of lymphocytes may generate high levels of autoantibodies or autoreactive T-cells. Malfunction of the mechanisms that maintain self-tolerance allows the immune system to inappropriately respond to cell's antigens, a condition that is referred to as **autoimmune response**.

The body must establish self-tolerance mechanisms, to distinguish between the self and non-self determinants, so as to avoid auto reactivity. Autoimmune diseases affect 5–7 per cent of the human population. Many such diseases are characterized by tissue destruction mediated directly by cells.

Autoimmune diseases in humans can be organ specific and/or systemic. The reactions of an autoimmune response cause serious damage to cells and organs. Sometimes antibodies cause the damage to self-cells or organs. In other cases T-cells are the culprit. A common form of autoimmunity is tissue injury by mechanisms similar to Type II hypersensitivity reactions. For example, in autoimmune haemolytic anaemia, antigens on red blood cells are recognized by autoantibodies, which cause the destruction of the blood cells resulting in anaemia.

An autoimmune response occurs only in the presence of self reactive T-cells and shows loss of tolerance to self-antigens because there is inadequate control of immune responses or regulatory failure. Loss of such tolerance may also be due to failure to maintain tolerance to self-antigens, and inability to distinguish between self and non-self antigens.

Thus, autoimmune response is an antigen-specific response of the immune system that may be either cell mediated or humoral. It is a normal response because it is directed against cell's antigens like cell surface receptors, erythrocytes surface proteins, basement membrane antigens of kidney glomeruli, hormone receptors, and nucleoproteins.

12.1.1 Self-reactive B- and T-cells Persist Even in Normal Subjects

The body contains large numbers of lymphocytes, which are potentially autoreactive. This is true of developing thymic T-cells (thymocytes) that failed to be eliminated by a subset of self-epitopes. Autoreactive T-cells are also present in normal individuals.

Many autoantigens, when injected with adjuvants, make autoantibodies in normal animals. This demonstrates the presence of autoreactive B-cells.

12.1.2 Chemokines in Autoimmune Diseases

One of the key processes that sets the stage for pathology associated with autoimmune response is activation and elicitation of leukocytes through cytokine driven networks. Chemokines are the mediators of the disease that provide the mechanism to elicit the appropriate population of leukocytes that are required to participate in an inflammatory response. Expression of chemokines and chemokine receptors is altered significantly during the evolution of multiple sclerosis. There are elevated levels of chemokines seen in multiple sclerosis and rheumatoid arthritis.

B-cells effector functions

A possible pathological consequence of autoantibody production is the development of in vivo immune complexes. Antigen-antibody complexes get localized in the capillary beds in several locations such as lung, brain and the skin. Damaging problems generally arise from immune complexes deposited in the glomeruli of kidneys.

T-cell effector functions

In human autoimmune diseases, the cytotoxic T-lymphocytes are the major effectors of damage to the organ localized autoimmune disorders. They include insulin-dependent diabetes, chronic thyroditis and multiple sclerosis. There is also evidence that shows an expansion of T-cell clones in inflamed rheumatoid arthritis joints.

Macrophage effector mechanism

Activated macrophages have important roles in cell mediated immunopathological reactions. Macrophages are source of cytokines, like IL-1 and IL-12 that enhance the production of CD4 T-cells. IL-12 enhances T-cell mediated immunity and the production of complement fixing isotypes of antibody. Interferon-γ is also an important product of activated macrophages. It participates in certain epithelial cell damage like thyroid follicle cells. It also enhances the activity of nitric acid synthase, which synthesizes nitric acid. Nitric acid along with reactive oxygen intermediates can bring about lot of tissue damage.

Dendritic Cells (DC)

These are specialized to initiate primary immune responses. DCs are also present in high numbers in the serum and synovial fluid or patients with rheumatoid arthritis. High levels of circulating DCs secrete pro inflammatory cytokines in multiple sclerosis. There is also evidence for dysfunctional DCs in autoimmunity. In individuals who are at risk of developing diabetes, dendritic cells express fewer co-stimulatory molecules and stimulate T-cells to a lesser extent. Mutations in genes affecting dendritic cell function had been directly related to autoimmunity.

Other effector mechanisms

Many mechanisms of innate immunity are involved in the damaging effects of autoimmune responses. NK-cells are important in the early stages of viral infection. These cells are found in autoimmune myocarditis and produce additional IFN-γ with potential immunopathological consequences.

Elevated levels of IFN in lupus erythematosus have been reported. Lupus patients have shown to have interferon inducing factor in the serum that can stimulate leukocytes from normal donors to produce large quantities of IFN.

12.1.3 Autoimmunity is Antigen Driven

T-cells are utterly pivotal for the development of autoimmune disease. The high affinity and somatic mutations, which are characteristic of the IgG part of antibodies, are dependent on the cooperative action of T_h-cells. In organ specific disorders, there is evidence for T-cells responding to antigens present in the organs under attack. Controls on the development of autoimmunity can be bypassed in a number of ways.

12.1.4 Physiological Autoimmunity

Red blood cells are removed from the bloodstream once they reach the end of their lifespan. As these cells age, the anion transporter glycoprotein called Band 3 is cleaved and a new epitope is exposed, which is recognized by a naturally occurring IgG autoantibody. Band 3 protein is also found on platelets, lymphocytes, neutrophils, hepatocytes and kidney cells. The cleavage of Band 3 in aged cells and the subsequent removal of those cells by antibodies may be mechanism for the elimination of old cells of many different types.

12.2 INDUCTION OF AUTOIMMUNITY

The precise mechanisms of breakdown of self-tolerance are not known for any autoimmune disease. However, several key events are likely to be involved. Induction of autoimmune disease is multifactorial and polygenic (Figure 12.1).

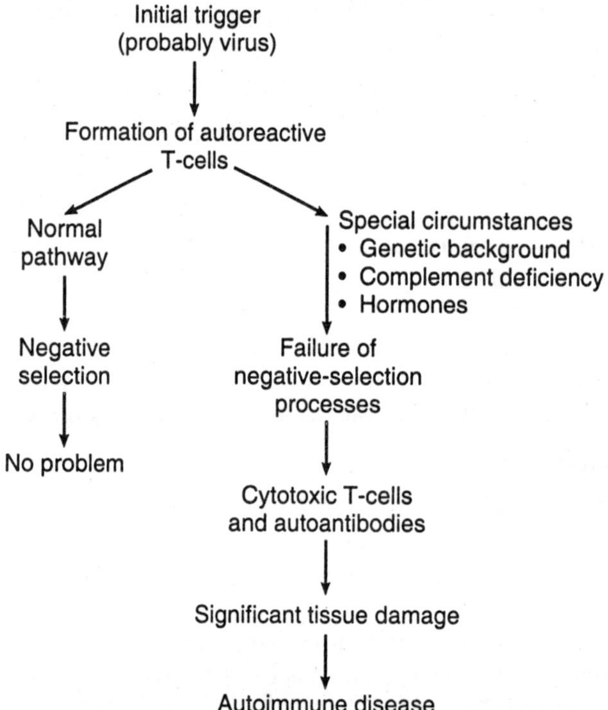

Figure 12.1 Some factors that influence the development of an autoimmune disease.

A variety of mechanisms have been proposed that can account for T-cell mediated generation of autoimmune diseases. It is quite likely that autoimmunity does not develop from a single event, but rather from a number of different events. There are multiple ways in which tolerance can break down, leading to autoimmunity.

These are as follows:

1. Thymic selection may be flawed in individuals with certain MHC genotypes.
2. Antigen cross-reactivity may activate self reactive T-cells (molecular mimicry).
3. Sequestered cells antigens may be released.
4. Cytokine production may be disturbed.
5. Immunoregulation may fail.

In some cases foreign antigens can directly stimulate auto reactive cells. Pre-existing defects in the target organ may also increase susceptibility to autoimmunity.

12.2.1 Exposure of Hidden Antigens or Release of Sequestered Antigens

Any tissue antigens that are sequestered from the circulation, and therefore are not seen by the developing T-cells will not induce self-tolerance. Exposure of mature T-cells to normal sequestered antigens at a later time may result in their activation. Some antigens exist within the body in places that are normally not visited by circulating lymphocytes. Antigens may be hidden in the central nervous system and the testes, where they are not usually visited by lymphocytes. Therefore, lymphocytes may not be completely tolerant to antigens in these organs. For example, sperms arise late in development and are sequestered from the circulation. After vasectomy, some sperm antigens are released into the circulation and can induce autoantibody formation in some men. If the brain or testes are injured, then the resulting leakage of blood vessels may permit antigens released by damaged cells to reach the bloodstream, encounter antigen sensitive cells, and stimulate an immune response. For example, after a myocardial infarction, autoantibodies may be produced against the intracellular components of cardiac muscle cells, such as mitochondria. Myelin Basic Protein (MBP) is an example of an antigen normally sequestered from the immune system by the blood brain barrier. Trauma to tissues following either an accident or a viral or bacterial infection might also release sequestered antigens into the circulation. Similarly the release of lens protein after eye damage or of heart muscle antigens after myocardial infarction leads to the formation of autoantibodies.

12.2.2 Formation of New Epitopes

The formation of autoantibodies may be provoked by the development of new epitopes on normal proteins. Examples are the Band 3 antigen on old red blood cells and rheumatoid factors.

12.2.3 Cross-reactivity with Microorganisms

Molecular mimicry is the sharing of epitopes between an infectious agent or parasite and its host. A pathogen may express a region of protein that resembles a particular self-component. When this occurs, an immune response directed against the invader may react with normal self-antigens and cause disease. The shared epitopes should be sufficiently similar for cross-reactivity to occur but sufficiently different for tolerance to be broken. *Trypanosoma cruzi* is a protozoan parasite, which possesses antigens that cross react with mammalian neurons and cardiac muscle. In infected individuals these antigens provoke autoantibody formation, which in turn provoke nervous disorders and heart disease. Epstein-Barr virus DNA polymerase cross-reacts with myelin basic protein and may be involved in the induction of multiple sclerosis.

12.2.4 Mimicry between MBP and Viral Peptides

Infections with certain viruses expressing epitopes that mimic sequestered self-components, such as myelin basic protein, may induce auto-immunity to those components. Molecular mimicry by cross-reactive microbial antigens can stimulate both autoreactive B- and T-cells. Post rabies encephalitis used to develop in some individuals who had received the rabies vaccine.

A number of viruses and bacteria possess antigenic determinants that are identical or similar to normal host cell components. Post rabies encephalitis in individuals who received rabies vaccine developed from rabbit brain cell culture. The vaccine prepared from this culture contains rapid brain cell antigen to which human body develops antibodies and activate T-cells leading to encephalitis. Mimicry is often seen between myelin basic protein peptides and viral peptides, such as measles virus, influenza, adenovirus. Molecular mimicry also occurs with heat shock proteins, which are produced by mammalian cells during extreme stressful conditions.

12.2.5 Loss of Control of Lymphocyte Responses

Most altered immune diseases probably occur as the result of the development of T- or B-cells that had previously been suppressed by the normal control mechanisms of the body. When mice are injected with rat red blood cells, they make antibodies to the rat cells and also develop a self-limiting and transient autoimmune response to their own red blood cells. Often autoimmune diseases are associated with lymphoid tumours; for example myasthenia gravis may be associated with a thymocyte tumour. In humans, there is a four-fold increase of rheumatoid diseases in patients with malignant lymphoid tumours. The stimulation of fas protein (CD95) on a cell surface causes a cell to undergo apoptosis. A defect in the fas protein may result in the development of autoimmunity by preventing negative selection of self reactive T- and B-cells.

12.3 THE FACTORS AFFECTING DEVELOPMENT OF AUTOIMMUNITY

There may be changes that occur in B- or T-lymphocytes or both populations, or there may be abnormal selection of lymphocytes during cell maturation. There might also be cross-reactions between shared antigens. Microorganisms that bear antigens that cross-react with self-antigens are able to activate self-reactive T-cells. The activated autoreactive T-cells are able to recognize and react against self-antigens and thereby produce an autoimmune response. Increased production of cytokines by T-cells is responsible for delayed type hypersensitivity reaction with tissue injury. Polyclonal stimulation of lymphocytes induces non-specific response of self-reactive lymphocytes and production of multiple autoantibodies, which are associated with non organ-specific autoimmune disease.

12.3.1 Viruses as Inducers of Autoimmunity

Viruses that infect lymphoid tissues may be capable of interfering with immunological control mechanisms and therefore, permit autoimmune disease to occur. Mice infected with certain reoviruses develop an autoimmune polyendocrine disease characterized by diabetes mellitus and retarded growth. These reovirus-infected mice develop antibodies against normal pituitary, pancreas, gastric mucosa, nuclei, glucagon, growth hormone, and insulin.

12.3.2 Polyclonal B-cell Activation

Viruses and bacteria can induce non-specific polyclonal B-cell activation . Gram-negative bacteria, cytomegalovirus, and Epstein-Barr Virus (EBV) are known polyclonal activators, inducing the proliferation of many clones of B-cells that express IgM in the absence of T_h-cells. If B-cells reactive to self-antigens are activated by the mechanism, autoantibodies can appear.

12.3.3 Genetic Factors and Autoimmune Disease

Genetic susceptibility factors can influence the development of autoimmune disease on several levels. Such factors include the MHC haplotype and polymorphism in genes involved in establishing self tolerance and immune regulation such as the T-cell Immunoglobulin and Mucin (TIM) domain containing family or Cytolytic T-Lymphocyte-associated Antigen 4 (CTLA-4). The MHC haplotype can influence susceptibility to a given autoimmune disease by enhancing the presentation of peptide epitopes in the periphery which results in increased T-cell activation, or by ineffective presentation of self-antigens in the thymus, which leads to more aggressive T-cells or a fewer number of regulatory T-cells.

Several genome scans carried out for patients with multiple sclerosis has revealed that HLA locus at 6p21 is most commonly involved in almost all cases.

Exposure of miners to silica is a precipitating factor in the autoimmune disease scleroderma, and mud resorts have been associated with autoimmune glomerulonephritis. Dietary iodine induces autoimmune thyroid disease.

Many genes influence the development of autoimmunity. In identical twins, when one gets multiple sclerosis, the other has a 30 per cent chance of developing the same disease. Genes for histocompatibility molecules, for peptide transporters, for T-cell antigen receptors, for complement components, and for immunoglobulins may play a role in autoimmunity. There are autoimmune diseases linked to MHC Class I and Class II molecules. Inappropriate expression of Class II MHC molecules may also lead to autoimmunity. The pancreatic beta cells of individuals with Insulin-Dependent Diabetes Mellitus (IDDM) express high levels of both Class I and Class II MHC molecules, whereas healthy beta cells express lower levels of Class I and do not express Class II MHC molecules at all. Similarly, thyroid acinar cells from those with Graves' disease express Class II MHC molecules on their membranes. This inappropriate expression of Class II MHC molecules , which are normally expressed only on antigen presenting cells, may serve to sensitize T_h-cells to peptides derived from the beta cells or thyroid cells, allowing activation of B-cells or T_C-cells or sensitization of T_{DTH}-cells against self-antigens.

Certain HLA haplotypes are predisposed to autoimmunity. Hashimoto's thyroditis tends to be associated more with DR5. Rheumatoid arthritis is associated with a nucleotide sequence that is common to DR1 and major subtypes of DR4. There are several genetic factors involved in the development of autoimmune diseases: there are genes predisposing individuals to develop autoimmunity, either organ specific or non-organ specific, and others determining the particular antigen or antigens involved. Individuals with HLA-DR4 are about six times more likely to develop rheumatoid arthritis, and those with HLA-DR53 have a tenfold increased susceptibility.

Genes also confer resistance to autoimmune diseases. Thus, resistance to juvenile diabetes correlates with the presence of the amino acid aspartate at position 57 of the HLA-DQβ chain. Other amino acids at this position such as serine, alanine, and valine are associated with increased susceptibility to the disease.

Autoimmune disease can occur in families. There is an undoubted familial incidence of autoimmunity. The autoimmunity within families often shows a bias towards organ-specific reactivity. There are genetically controlled factors, which tend to select the organ that is mainly affected. Relatives of Hashimoto patients and relatives of pernicious anaemia patients, both, have a higher incidence and titre of thyroid autoantibodies, whereas relatives of pernicious anaemia have a higher frequency of gastric autoantibodies. Thus, there are genetic factors, which differentially select the stomach as the target within this group of organ specific autoimmune disorders.

Since certain genes have been implicated in the development of autoimmunity, heredity is also a possible factor for autoimmunity. Individuals who have inherited a particular major histocompatibility complex of genes like HLA-DR4 have a higher probability of developing an autoimmune skin disease known as **pemphis vulgaris**.

12.3.4 Age and Sex

The frequency of autoimmune diseases increases with age (Figure 12.2). They are also more frequent in women than in men, perhaps because different cells in females express different X chromosomes (Lyon effect); the resulting cellular mosaicism might increase opportunities for a breakdown in self-tolerance. Susceptibility to many autoimmune diseases also differs between the two sexes. Hashimoto's thyroditis, multiple sclerosis, systemic lupus erythematosus and rheumatoid arthritis preferentially affect women. 90 per cent or more of patients with SLE are women. A possible explanation for this is that estrogens significantly enhance IFN-γ production and IFN-γ plays a major role in the pathogenesis of tissue destruction in these diseases by enhancing local MHC expression and activating macrophages.

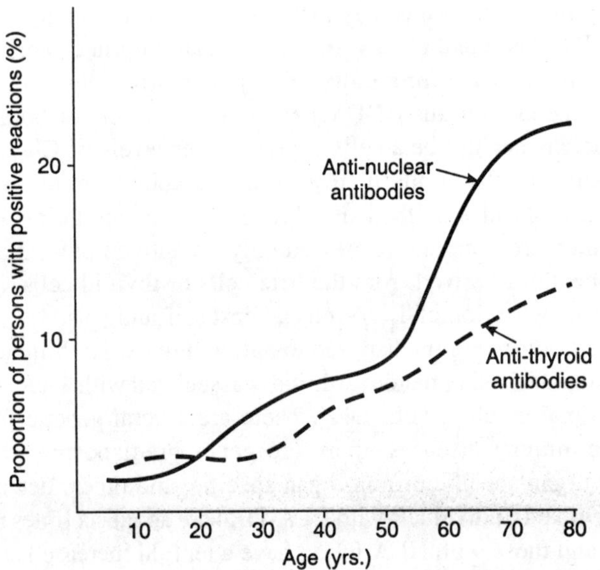

Figure 12.2 Increasing incidence of autoantibodies with age in general population.

12.4 CLINICAL ASSOCIATIONS BETWEEN AUTOIMMUNE DISEASES

Very often more than one autoimmunity disease occurs simultaneously in the same individual. Autoimmune thyroditis is commonly associated with autoimmunity to gastric parietal cells and adrenal cortex. Half of the patients with pernicious anaemia have antithyroid antibodies. The simultaneous occurrence of multiple autoimmune disorders is even more marked in SLE, in which haemolytic anaemia, thrombocytopenia and rheumatoid arthritis are all reported.

Autoimmunity may also develop as a consequence of certain alterations that occur locally during inflammation or tissue injury like release of previously removed self-antigens, and alterations in the structure of self-antigens. Environmental factors like drugs and ultraviolet radiation may produce alterations in certain self-antigens. The immune system then recognizes the altered self-antigens as non-self and generates an autoimmune response.

Autoantibodies are designated according to the specific antigens that induced their production and are classified as *organ specific* and *organ non-specific antibodies*. The organ specific autoantibodies are directed against antigens that are specific for a particular organ and include cell surface receptors, biologically active molecules like hormones and specific tissue/organ antigens.

In organ specific autoimmune diseases the immune response is directed against a specific antigen corresponding to a single organ. As a result of the attack by humoral or cell mediated mechanism the organ may suffer severe damage or the function of the organ may be enhanced due to autoimmune attack resulting in hyperfunctioning, or the autoantibodies may block the function performed by the organ and result in its failure.

In a lot of immune diseases, autoantibodies act as agonists and bind to receptors of hormones and bring about similar effect. This results in overproduction of a particular cellular product or increase in cell growth. Conversely autoantibodies may act as antagonists and block the receptor from actual hormonal or neurotransmitter action, thereby impairing the cellular function.

In systemic autoimmune diseases the autoantibodies are directed against a broad range of antigen and target organs. This is due to defective immune regulation of T- and B-lymphocytes, which leads to wide tissue damage from cell mediated and humoral antibodies as well as immune complex mediated response.

The organ non-specific autoantibodies do not show specificity for any one organ, but may be directed against multiple organs.

12.5 PATHOGENESIS

When autoantibodies are found in association with the particular disease there are three possible inferences.

1. The autoimmunity is responsible for producing the lesions of the disease.
2. There is a disease process which, through the production of tissue damage, leads to the development of autoantibodies.
3. There is a factor which produces both the lesions and the autoimmunity.

Autoantibodies secondary to a lesion are sometimes found. For example, cardiac autoantibodies may develop after a myocardial infarction. In most diseases associated with autoimmunity, evidence supports that the autoimmune process produces the lesions.

12.6 SOME SELECTED AUTOIMMUNE DISEASES

Autoimmune thyroditis

In humans, two major forms occur. In **Hashimoto's thyroditis**, the autoantibodies formed against thyroid peroxidase can activate complement. These antibodies, together with complement, can destroy thyroid cells. Hashimoto's thyroditis is frequently seen in middle-aged women and the individual produces autoantibodies and sensitized T_{DTH}-cells specific towards thyroid antigens. The ensuing inflammatory response causes a goiter or visible enlargement of the thyroid gland.

Thyrotoxicosis or Graves' disease is associated with excessive thyroid activity. The autoantibodies formed in thyrotoxicosis bind to the same receptor as Thyroid Stimulating Factor (TSH) and instead of destroying cells, enhance thyroid activity.

Myasthenia gravis

This is a disease of the skeletal muscle characterized by abnormal fatigue and extreme weakness after relatively mild exercise. The muscles holding up the eyelids tire, making it difficult for the affected individuals to keep their eyes open. Early signs of this disease include drooping eyelids. The muscles of the oesophagus relax and make swallowing difficult. Myasthenia gravis results from destruction of acetylcholine receptors on the motor end plates of striated muscle. This destruction is caused by IgG autoantibodies against acetylcholine receptor protein. These autoantibodies may block binding sites, can activate complement, and they can accelerate endocytosis and degradation of receptors. As a result, the number of acetylcholine receptors is reduced, and repeating the stimulus is ineffective since all available receptors are saturated with acetylcholine. In pregnant women, suffering from myasthenia gravis, antibodies to acetylcholine receptors cross the placenta into the foetus and cause transient muscle weakness in the newborn baby (Figure 12.3).

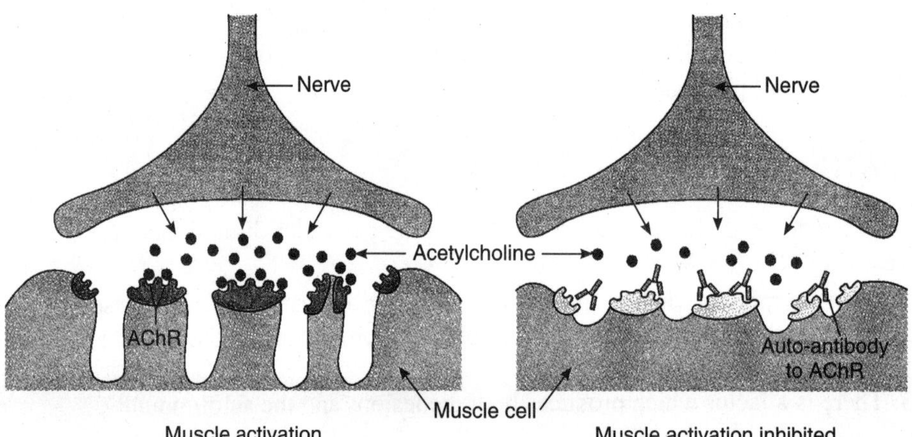

Figure 12.3 In myasthenia gravis, binding of autoantibodies to acetylcholine receptors (right) blocks the normal binding of acetylcholine and subsequent muscle activation (left).

Systemic lupus erythematosus

This is a generalized immunological disorder. About 90 per cent of cases occur in women and result from a loss of control of the B-cell system. SLE is a systemic autoimmune disease that appears in women between 20 and 40 years of age; the ratio of male to female patients is 10 : 1. Fever, weakness, arthritis, skin rashes, pleurisy, and kidney dysfunction characterize SLE.

The affected individuals make autoantibodies against a range of antigens found in normal organs and tissues. These multiple autoantibodies cause a wide spectrum of pathological lesions and clinical manifestations. One consistent feature of SLE is the development of autoantibodies against nucleic acids. These autoantibodies can combine with free DNA to form DNA-anti-DNA immune complexes, which may be deposited in glomeruli causing a membranous glomerulonephritis. The complexes may also be deposited in the walls of arteries causing tissue destruction and fibrosis, or in joint synovia where they provoke arthritis. When antinuclear antibodies bind to the nuclei of degenerating cells, these opsonized nuclei may be phagocytosed forming structures known as **lupus erythematosus cells**. Autoantibodies to red blood cells cause an antiglobulin positive haemolytic anaemia. Antibodies to platelets give rise to an immunologically mediated thrombocytopenia. There is evidence that suggests that viral or microbial infections may be responsible for initiation of the condition. Many SLE patients are defective in their ability to remove immune complexes from their circulation.

Rheumatoid arthritis

This is a common, crippling disease in humans and involves the joints and other body systems. It occurs most often in women from 40 to 60 years of age. There is a chronic inflammation of the joints. The disease starts when B-lymphocytes in the synovium are activated and produce autoantibodies including the rheumatoid factor. The patients produce a group of autoantibodies (of the IgM class) called **rheumatoid factors** that are reactive with IgG. Such autoantibodies bind to normal circulating IgG, forming IgM-IgG complexes that are deposited in the joints. These complexes can activate the complement cascade, resulting in Type III hypersensitive reactions, which leads to chronic inflammation. Some of these antibodies may be directed against collagen. There is erosion of articular cartilage and the neighbouring bony structures too.

Organ specific autoimmune diseases

These are diseases mediated by direct cellular damage. Such diseases are caused when lymphocytes or antibodies bind to cell membrane antigens causing cellular lysis and/or an inflammatory response in the affected organ. The damaged cellular structure is gradually replaced by connective tissue and the function of the organ declines.

Autoimmune anaemias

They include pernicious anaemia, autoimmune haemolytic anaemia, and drug induced haemolytic anaemia. In pernicious anaemia, the absence of sufficient vitamin B12, which is necessary for proper haematopoiesis, leads to a decline in the number of functional mature red blood cells. Pernicious anaemia is treated with injections of vitamin B12. To be adsorbed, Vitamin B12 must be first associated with the protein called **intrinsic factor** and then get transported across the intestinal mucosa. Plasma cells in the gastric mucosa of patients with pernicious anaemia secrete antibodies against intrinsic factor into the lumen of the stomach. An individual with autoimmune haemolytic anaemia makes autoantibodies against RBC antigens, triggering complement-mediated lysis or antibody mediated opsonization and phagocytosis of red blood cells.

Goodpasture's syndrome

Here, auto antibodies specific for certain basement-membrane antigens bind to the basement membranes of the kidney glomeruli and the alveoli of the lungs. Complement activation leads to direct cellular damage and an ensuing inflammatory response mediated by a buildup of complement split products. There is progressive kidney damage and pulmonary haemorrhage; death ensues within several months of the onset of symptoms.

Insulin-Dependent Diabetes Mellitus (IDDM)

This is caused by an autoimmune attack on the pancreas, which is directed against insulin producing cells. The autoimmune attack destroys the beta cells. There is a decreased production of insulin leading to increased levels of blood glucose. The destruction of beta cells results in serious metabolic problems that include ketoacidosis and increased production of urine, followed by atherosclerotic vascular lesions, renal failure and blindness.

Male infertility

An example of autoimmune disease is seen in the rare cases of male infertility where antibodies to spermatozoa leads to clumping of spermatozoa in the semen, either by their heads or by their tails.

Diseases mediated by stimulating or blocked autoantibodies

In some autoimmune diseases, antibodies act as agonists. They bind to hormone receptors instead of the normal ligand and stimulate inappropriate activity resulting in over production of mediators or an increase in cell growth. In other autoimmune conditions, autoantibodies bind to hormone receptors, but act as antagonists, blocking receptor function.

Graves' disease

This disease produces autoantibodies to the receptor for Thyroid Stimulating Hormone (TSH). Binding of the autoantibodies to the receptor mimics the normal action of TSH and activates adenylate cyclase. This results in uncontrolled production of the thyroid hormones.

Multiple sclerosis

This is an autoimmune disease of the central nervous system. In infected individuals, the myelin sheaths that surround the nerve axons in the white matter of the brain are progressively destroyed by immunological attack. The effects include numbness in the limbs, and going to paralysis or loss of vision. People between the ages 20 and 40 are usually affected. Autoreactive T-cells are produced in this disease that participate in the formation of inflammatory lesions along the myelin sheath of nerve fibres causing a number of neurologic dysfunctions. The nerve transmission is disrupted. Patients lose their ability to control muscle function.

12.7 THERAPIES

Current therapies for autoimmune diseases are aimed at reducing symptoms to provide the patient with an acceptable quality of life (Figure 12.4). These treatments provide non-specific suppression of the immune system and thus, do not distinguish between a pathologic autoimmune response and a protective immune response.

In organ specific autoimmune disorders, the symptoms can be corrected by metabolic control. Hypothyroidism can be controlled by administration of thyroxine; and thyrotoxicosis by antithyroid drugs. In myasthenia gravis, metabolic correction is through administration of cholinesterase inhibitors. Where function is lost and cannot be substituted by hormones, as in chronic rheumatoid arthritis, tissue grafts or mechanical substitutes may be appropriate. Conventional immunosuppressive therapy with antimitotic drugs can also be used. Immunosuppressive drugs (e.g. corticosteroids, azathioprine, and cyclophosphamide) are often given for slowing the proliferation of lymphocytes. A more selective approach uses cyclosporin A or FK506 to treat autoimmunity.

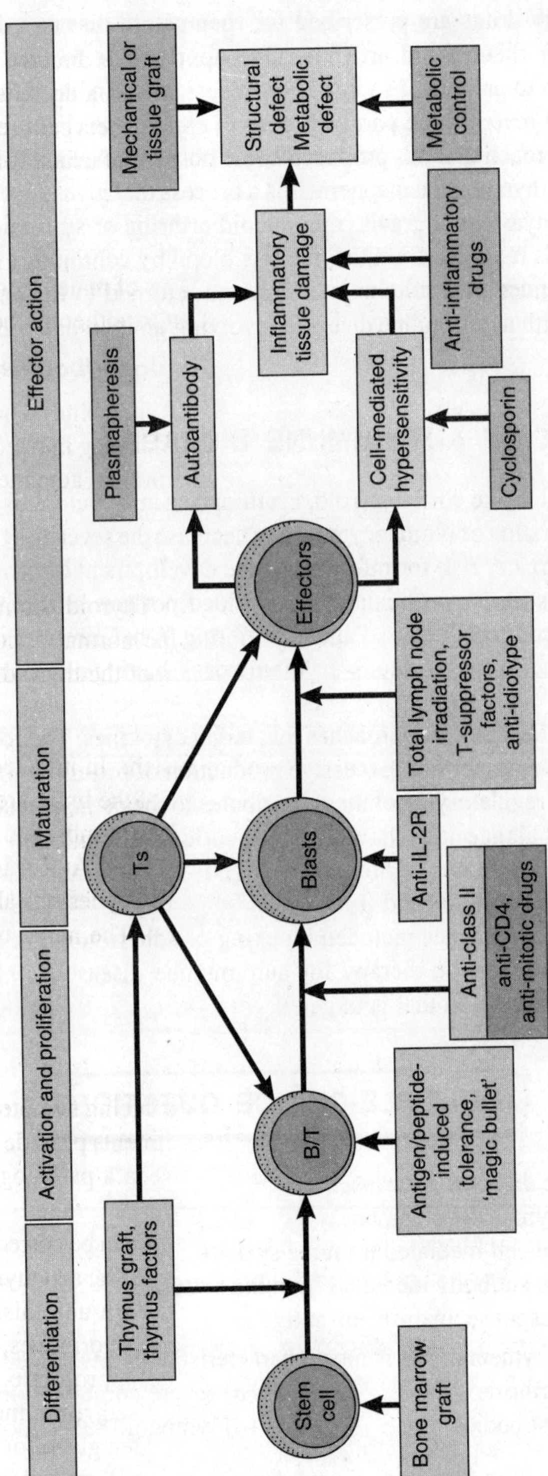

Figure 12.4 Current and potential treatment of autoimmune disease.

Anti-inflammatory drugs are prescribed for rheumatoid diseases. Methotrexate is the best available treatment for rheumatoid arthritis. One approach is to feed large quantities of the appropriate autoantigen to patients. Feeding myelin basic protein has induced tolerance to myelin proteins in animals and reversed the paralysis seen in experimental autoimmune encephalitis.

A therapeutic approach that has produced some positive results in some cases of myasthenia gravis is the removal of thymus. Plasmapheresis is a process that gives short-term benefit to patients with Graves' disease, myasthenia gravis, rheumatoid arthritis or systemic lupus erythromatosus. In this process plasma is removed from the patients blood by continuous flow centrifugation. The RBCs are then resuspended in a suitable medium and returned to the patient. Plasmapheresis is beneficial to patients with autoimmune diseases involving antigen-antibody complexes, which are removed with plasma.

12.8 TREATMENT OF AUTOIMMUNE DISEASES

Immunosuppressive drugs like corticosteroid, azathioprine and cyclophosphamide are often given to slow down the proliferation of lymphocytes. They decrease the severity of autoimmune symptoms, but puts the patient at greater risk for infection or the development of cancer. Removal of thymus in autoimmune diseases like myasthenia gravis yielded positive results, because in this disease thymic hyperplasia is a common feature. Patients suffering from Graves' disease, myasthenia gravis, rheumatoid arthritis and systemic lupus erythematosus have gained short-term benefits from plasmapheresis.

There are certain therapeutic approaches that target cytokines. The role of different cytokines in autoimmune diseases is important. Excessive production of pro-inflammatory cytokines as well as a relative shortage of regulatory cytokines contributes to the pathogenesis of diseases. Restoring the optimal cytokine balance may have a therapeutic value and can be achieved by either neutralization of pro-inflammatory cytokines like TNF-α, IL-12 and IL-6, or by inducing anti-inflammatory cytokines like IL-10 and IFN-β.

Other therapeutic approaches included blocking of adhesion molecules as well as blocking co-stimulatory pathways. Genetic therapy for autoimmune diseases is also done. There can be reconstitution of gene expression that is targeted.

MULTIPLE-CHOICE QUESTIONS

1. The autoimmune diseases are caused by:
   (a) a defect in thymus development
   (b) a defect in the cell mediated immune system
   (c) a defect in the antibody mediated immune system
   (d) an immune response against self-antigens
2. Systemic lupus erythematosus is most characteristically associated with:
   (a) rheumatoid arthritis
   (b) severe combine immunodeficiency
   (c) antinuclear antibodies
   (d) lymphoid cell tumours

3. Aged red cells are removed by the autoantibodies against:
   (a) band 3 protein
   (b) type O blood group
   (c) complement
   (d) idiotypes
4. Rheumatoid factor is an:
   (a) antinuclear antibody
   (b) antibody to complement
   (c) antibody to immunoglobulin
   (d) antibody to synovial membranes
5. Some autoimmune diseases may result from cross-reactions between organisms and tissues. This phenomenon is called:
   (a) molecular mimicry
   (b) antigenic variation
   (c) tissues sensitization
   (d) clonal anergy
6. Thyrotoxicosis results from autoimmune attack against the receptor for:
   (a) acetylcholine
   (b) insulin
   (c) thymic hormone
   (d) thyroid stimulating hormone

REVIEW QUESTIONS

12.1 Discuss the organ specific autoimmune diseases.

12.2 Discuss the systemic autoimmune diseases.

12.3 Write notes on:
   (i) Myasthenia gravis
   (ii) Systemic lupus erythematosus
   (iii) Rheumatoid arthritis
   (iv) Plasmapheresis

13

IMMUNODEFICIENCY

13.1 IMMUNODEFICIENCY

When the immune system errs by failing to protect the host from disease causing agents or from malignant cells, it is termed immunodeficiency.

Immunodeficiency diseases arise from various abnormalities in the immune system, which result from congenital effects or due to acquired defects. Congenital immunodeficiencies are due to abnormalities in certain genes that control haematopoiesis or development of leukocytes or immunoglobulin production. Acquired immunodeficiencies are due to prolonged viral or bacterial infections, malnutrition or due to drug treatment.

Immunodeficiency can be a defect or deficiency in any one or more of the main components of the immune system (example B-lymphocytes, T-lymphocytes, phagocytic cells or complement components) which may result in an abnormally low immune response. The deficiency may affect natural immune responses (e.g. defect in phagocytosis or complement system) or specific immune responses (e.g. abnormality in function, development or activation of cells of the immune system). Sometimes both natural and specific immune responses may be involved in immunodeficiency. Immunodeficiency disorders have been classified into two general categories, primary and secondary.

13.1.1 Primary Immunodeficiencies

These are also known as congenital immunodeficiency or hereditary immunodeficiency, and are caused by a genetic defect that is inherited and becomes evident as an increased susceptibility to infections in early childhood.

In such a condition, the defect is present at birth although it may not manifest itself until later in life. A primary immunodeficiency may affect either adaptive or innate immune functions. Deficiencies of the specific components of adaptive immunity, such as T- or B-cells, are thus differentiated from immunodeficiencies in which the non-specific mediators of innate immunity, such as macrophage or complement, are impaired. Most defects that lead to immunodeficiencies affect either lymphoid or myeloid function. The lymphoid cell disorders may affect T-cells, B-cells, or both B- and T-cells. The myeloid cell disorders affect phagocytic function. Most of the primary immunodeficiencies are inherited. There are immunodeficiencies that stem from developmental defects that impair proper function of an organ of the immune system. The consequences of primary immunodeficiency depend on the number and type of immune system components involved.

Treatment of these disorders is limited to managing the infection and replacing abnormal components through transplantation.

13.1.2 Secondary Immunodeficiencies

These are also known as acquired immunodeficiencies and develop later in life. They are secondary or subsequent to infection and cancers, as well as to treatment with immunosuppressive drugs, or to diabetes, burns or malnutrition. Thus, secondary or acquired immunodeficiency is the loss of immune function and results from exposure to various agents.

Immunodeficiency disorders may also be defined according to the component of the immune system that shows that defect. For instance we have B-cell deficiencies, T-cell deficiencies, combined B-cell and T-cell deficiencies, defect in phagocytic function and complement deficiencies.

13.2 DEFECTS IN THE LYMPHOID LINEAGE

Lymphoid immunodeficiencies may involve B-cells, T-cells or both of these lineages. The combined form of lymphoid immunodeficiency that affects both lineages, are generally lethal within the first few years of life.

Defects in the humoral system are associated primarily with infections by encapsulated bacteria, while defects in the cell-mediated system are associated with increased susceptibility to viral, protozoan, and fungal infections. Infections with viruses that are rarely pathogenic for normal individual may be life threatening for those with impaired cell mediated immunity. The immunodeficiencies that affect lymphoid function have a common inability to mount or sustain a complete immune response against specific agents. A variety of failures can lead to such immunodeficiency.

13.2.1 B-cell Immunodeficiency Disorders

Any deficiency in antibody production will adversely affect the individuals' ability to protect against infections. These may be detected as lower serum immunoglobulin levels. B-cell immunodeficiency disorders make up various diseases that range from complete absence of mature recirculating B-cells, plasma cells, and immunoglobulin to the selective absence of only certain classes of immunoglobulins. These heredity defects may be at the B-lymphocyte maturation stage (pre-B-cell to B-cell) or in the B-lymphocyte response to an antigen, such as is seen in heavy chain class switching defect. Abnormality in helper ($CD4^+$) T-lymphocytes may also adversely affect antibody production.

Failure in any stage of B-cell formation or maturation or gene expression in B-cells would lead to **humoral deficiencies**. People having B-cell disorders suffer from recurrent bacterial infections, but display normal immune response against viruses and fungal infections because T-cell immune response in these patients is normal.

X-Linked Agammaglobulinaemia (XLA). Children afflicted with this disease have recurrent bacterial infections at about six months of age, by the time the passively acquired maternal antibody levels has fallen. The gene for this defect is present on the long arm of the X chromosome. This disease is prevalent in male children. XLA causes failure of the pre B-cells to mature into B-cells. People having XLA show extremely low concentration of serum immunoglobulins. In such persons the pre B-cells have immunoglobulin heavy chain genes rearranged, but the light chain genes remain in the germline configuration.

The extent of B-cell immunodeficiency may be limited to a single class deficiency (e.g. IgA or IgG subclass deficiency) or it may involve a complete absence of B-lymphocytes and serum immunoglobulins (e.g. X-linked agammaglobulinemia in male infants, also known as **Bruton's agammaglobulinemia**). This is a relatively rare X-chromosome linked defect, in which the pre-B-cells are present, but because of a defective (mutated) gene located on the X-chromosome, the pre-B-cells cannot mature. Thus, B-cells and lymphoid tissue (e.g. tonsils) are absent resulting in severe depression or absence of all classes of immunoglobulins. A female (XX) who carries a defective gene shows no abnormality because of the presence of another normal X-chromosome. However, males (XY) who inherit the abnormal X-chromosome show absence of antibodies in their serum. During infancy, protection from recurrent bacterial infection is provided by maternal IgG. After depletion of maternal IgG, administered gamma globulin maintains the protection.

13.2.2 T-cell Immunodeficiency Disorders

Several immunodeficiency syndromes are because of the failure of the thymus to undergo normal development. There is a profound effect on T-cell function; all populations of T-cells including helper, cytolytic, and regulatory varieties are affected. Immunity to viruses and fungi is especially compromised in those suffering from these conditions.

Cell mediated immune responses are affected in these disorders. Humoral immune response may also be affected because of reduced help from $CD4^+$ T-cells.

There is a decreased number of total blood T-lymphocytes and a lowered proliferation of the cells in response to antigens or to T-cell activators. T-cell immunodeficiency disorders may be as follows:

(i) Primary or congenital T-cell immunodeficiency.
(ii) Secondary or acquired T-cell immunodeficiency.

Primary T-cell immunodeficiency

This may be the outcome of defective maturation of T-lymphocytes caused by an abnormal development of the thymus. It is characterized by a reduction or absence of blood T-cells. The deficiency may be caused by (i) defective surface receptors, (ii) defective cytokine production, or (iii) impaired cell-to-cell communication. DiGeorge syndrome, or congenital thymic aplasia, is a developmental defect that is associated with the deletion of a region on chromosome 22 in the embryo, causing immunodeficiency along with characteristic facial abnormalities, hypo-parathyroidism, and congenital heart disease. The immune defect includes a profound depression of T-cell numbers and absence of T-cell responses. Although B-cells are present in normal numbers, affected individuals do not produce antibody in response to immunization with specific antigens because of the absence of helper T-lymphocytes. Thus, an abnormal humoral response is also observed. A congenital T-cell deficiency is characterized by a reduction or absence of T-lymphocytes in the blood, which results in an impaired cell mediated immune response. The severity of the immunodeficiency depends on the degree of impairment in the development of the thymus. DiGeorge syndrome (congenital thymic aplasia) is characterized by the absence of thymus and is associated with hypothyroidism, cardiovascular anomalies and increased incidence of infections. People afflicted with this disease have an effective humoral immune response against natural infection, but are highly susceptible to viral and fungal infections. There is a severe decrease in the total number of T-cells.

Thymic hypoplasia, or the Nezelof syndrome is an inherited disorder. Here a vestigial thymus is unable to serve its function in T-cell development. In some patients, B-cells are normal, whereas in others a B-cell deficiency is secondary to the T-cell defect. Affected individuals suffer from chronic diarrhoea, viral and fungal infections, and a general failure to thrive.

Secondary T-cell immunodeficiency

This is associated with an increased susceptibility to malignant tumours and infections that are caused by a variety of opportunistic microorganisms. Acquired T-cell deficiencies may be caused by a loss of $CD4^+$ T-cell functions. Acquired immunodeficiency syndrome is a secondary immunodeficiency condition induced by HIV infection of $CD4^+$ T-cells, macrophages, and dendritic cells within the lymph nodes. There is a decreased number of $CD4^+$ T-cells and an impairment of the antigen-specific $CD4^+$ function. This is accompanied by lowered production of IL-2, affecting the activation of T-cell subsets and cytotoxic T-cells.

13.2.3 Combined B-cell and T-cell Immunodeficiency Disorders

These affect both B- and T-lymphocytes. The immunodeficiency may be caused by primary lymphoid deficiencies or may be associated with other hereditary disorders. The conditions that are responsible for T- and B-cell combined immunodeficiencies may be

(i) Defect in maturation of T- and B-lymphocytes, causing reduction in number of T- and B-cells and serum immunoglobulins
(ii) Enzyme deficiency, leading to accumulation of toxic substances
(iii) Abnormal maturation of bone marrow stem cells, resulting in reduction in number of T- and B-cells, other cells, and serum immunoglobulin levels.

In Severe Combined Immunodeficiency Disease (SCID), there is a defect in B- and T-cell development from bone marrow stem cells, decrease in lymphocytes (lymphopenia), and a deficient humoral and cell-mediated immune response. SCID is associated with increased susceptibility to bacterial, fungal and protozoan infections due to insufficient number of B- and T-cells. Over half of the SCID cases are caused by X-linked recessive defects. The defect lies in the gene encoding the γ subunit of a number of cytokine receptors.

Several forms of human SCID arise from different genetic defects. The first described form of human SCID was an autosomal recessive disorder. The gene responsible is not encoded on the X- or Y-chromosome and only homozygotes are affected. There is also an X-linked form occurring predominantly in males, called primary lymphopenic immunologic deficiency, which is characterized by absence of immunoglobulin in the circulation.

Clinically SCID is characterized by a very low number of circulating lymphocytes. There is a failure to mount immune responses mediated by T-cells. The thymus does not develop and the few circulating T-cells cannot proliferate in response to antigens. Myeloid and erythroid cells appear normal in number and function, indicating that only lymphoid cells are depleted in SCID.

SCID results in severe recurrent infections and is usually fatal in the early years of life. SCID infants suffer from chronic diarrhoea, pneumonia, and skin, mouth, and throat lesions as well as a host of other opportunistic infections. Avoiding all potentially harmful microorganisms can prolong the life span of a SCID patient. Survival beyond the first year without treatment is rare. Bone marrow transplantation cures these infants.

Individuals with this inheritable disease are susceptible to microbial infections, such as *Candida*, *Cytomegalovirus*, and the *Pneumocystis carinii*. A defect leading to general failure of immunity similar to SCID is failure to transcribe the genes that encode Class II MHC molecules. Without these molecules, the patient's lymphocytes cannot participate in cellular interactions with T helper-cells. This immunodeficiency is called **bare-lymphocyte syndrome.**

13.3 OTHER IMMUNODEFICIENCY DISORDERS

These may occur following certain therapeutic procedures or because of an abnormality in phagocytic cells and complement components. These deficiencies may be:

(i) Secondary to chemotherapy/radiation
(ii) Caused by deliberate immunosuppression
(iii) Resulting from defect in phagocytic function
(iv) Caused by congenital defects in complement components

There are no cures for immunodeficiency disorders. However, there are several treatment possibilities which include:

(i) Replacement of a missing protein
(ii) Replacement of a missing cell type or lineage
(iii) Replacement of a missing or defective gene

13.3.1 Phagocytes Deficiencies

Phagocytes constitute an important arm of the immune system as they play a major role in natural or innate immunity. Defects in the phagocytes defence mechanisms result in immunodeficiency, which may be because of reduction in the number of phagocytes or from reduction in their function. Individuals suffering from phagocytic defects constantly suffer from bacterial or fungal infections ranging from mild skin infections to life threatening systemic infections with various complications.

Reduction in neutrophil counts is called **neutropenia** or granulocytopenia. Here the peripheral blood neutrophil count goes below 1500/µl or sometimes there is a complete absence of neutrophils which is termed agranulocytosis. This may happen because of a congenital defect or an acquired one. An autoimmune disease called Sjogren's syndrome or systemic lupus erythematosus leads to neutropenia because the autoantibodies act on neutrophils and destroy them.

Defective phagocytic functions which include adherance defects, chemotactic defects and killing defects also lead to autoimmune disorders.

Defects in the myeloid lineage include reduction in neutrophil count, Chronic Granulomatous Disease (CGD), and leukocyte adhesion deficiency.

Pooled human gamma globulin given intravenously or subcutaneously protects against recurrent infection in many types of immunodeficiency. It is now possible to clone the genes that code other immunologically important proteins such as cytokines, and to express these genes in vitro. The administration of recombinant IFN-γ has proven effective for patients with Chronic Granulomatous Disease (CGD), and the use of recombinant IL-2 may help to restore immune function in AIDS patients.

Replacement of stem cells with those from an immunocompetent donor allows development of a functional immune system. High rates of success are seen when there is an HLA-identical donor. Because of the recent nature of this therapy, it is not known whether transplantation cures the immunodeficiency permanently.

If a single gene defect is identified, as in adenosine deaminase deficiency or CGD, replacement of the defective gene may be a treatment option.

13.4 AIDS AND OTHER ACQUIRED OR SECONDARY IMMUNODEFICIENCIES

A variety of defects in the immune system give rise to immunodeficiency. In addition to the primary immunodeficiencies, there are also acquired, or secondary, immunodeficiencies like hypogammaglobulinemia. Another form of secondary immunodeficiency, known as **agent-induced immunodeficiency**, results from exposure to any of a number of chemical and biological agents that induce an immunodeficient state. Certain of these are drugs used to fight autoimmune diseases such as rheumatoid arthritis or lupus erythematosus. Corticosteroids, which are commonly used for autoimmune disorders, interfere with the immune response in order to relieve disease symptoms and induce an immunodeficiency state in those treated. Similarly, a state of immunodeficiency is deliberately induced in transplantation patients who are given immunosuppressive drugs, in order to blunt the attack of the immune system on the transplanted organs. There are recent efforts to use more specific means of inducing tolerance to allografts to circumvent the unwanted side effects of general immunosuppression.

13.4.1 Human Immunodeficiency Virus (HIV)

In 1983, Luc Montagnier and his team at the Pasteur Institute in Paris isolated a retrovirus from the lymph node biopsy of an AIDS patient. The retrovirus was named human immunodeficiency virus. In 1986, an antigenic variant of HIV was discovered; the original virus was designated HIV-1 and the gradient was called HIV-2. Both these viruses are genetically related to simian immunodeficiency virus. Most HIV-2 strains are less pathogenic and spread slower than HIV-1.

The HIV infection of a target cell involves two steps: binding of the virions to the target cells followed by fusion of the viral membrane envelop with the plasma membranc of the target cells. gp120 helps in the binding of the virion to the target cell,while gp41 helps in the fusion. Once the virus makes its entry into the target cells, the viral genome, RNA, is uncoated and transcribed into DNA, which is then integrated into the host cell genome forming the *provirus*, which may remain in this latent state or may be activated and transcribed into viral proteins immediately. The proviral state is facilitated by integrase. The integrated viral DNA is permanently associated with host cell DNA and is passed on to the daughter cells on cell division. As long as the provirus remains in the latent state, the viral genes are not expressed and so the virus remains hidden from the host immune system for a long time.

The primary cause of AIDS or acquired immunodeficiency syndrome is HIV-1 which has been isolated from blood, bone marrow, lymph nodes, spleen, plasma, semen, and saliva of infected individuals.

This deficiency syndrome is novel because a type of virus called a retrovirus causes it. Retroviruses carry their genetic information in the form of RNA. When the virus enters a cell, the RNA is reverse transcribed to DNA by a virally encoded enzyme, Reverse Transcriptase (RT). RT reverses the normal transcription process and makes a DNA copy of the viral RNA genome. This copy which is called a **provirus**, is integrated into the cell genome and is replicated along with the cell DNA. When the provirus is expressed it forms new virions and the cell lyses. Alternatively, the provirus may remain latent in the cell until some regulatory signal starts the expression process.

The HIV-1 is seen as a spherical virus containing a dense core surrounded by a lipoprotein envelope that is acquired as the virus buds from the surface of infected cell.

In the centre of the virion are two strands of RNA, each containing about 10,000 nucleotides. Attached to these strands are molecules of the reverse transcriptase, a protease, a ribonuclease and an integrase (an endonuclease that helps to integrate viral DNA into host cell DNA). All these are contained within a capsid in the shape of a hollow truncated cone. The capsid is made of a protein p24. Surrounding the capsid is the matrix protein p17. The entire mass is covered by an envelope consisting of a lipid bilayer derived from the outer cell membrane of infected cell. It may therefore also carry cell surface glycoproteins such as MHC molecules. Many spikes formed by two glycoproteins are present on the viral envelope. One gp41 is embedded in the membrane while four molecules of the other glycoprotein gp120 form the spike.

HIV genome is well characterized. The viral RNA has two flanking Long Terminal Repeat (LTR) sequences that act as binding sites for host transcription factors (Figure 13.1).

There are *three structural genes.*

Gag gene: It codes for a 53-kDa precursor that forms the following nucleocapsid proteins:

- p17 which forms outer core protein layer.
- p24 which forms the inner core protein layer.
- p9 which is a component of nucleotide core.
- p7 that binds directly to genomic RNA.

Env gene: It codes for a 160-kDa precursor that forms the following envelop glycoproteins:

- gp41 that is a transmembrane protein associated with gp120 and required for fusion.
- gp120 which protrudes from envelop and binds CD4.

Pol gene: It codes for a precursor that forms the following enzymes:

- p64 which has reverse transcriptase and RNase activity.
- p51 which has reverse transcriptase activity.
- p10 is a protease that cleaves gag precursor.
- p32 is integrase.

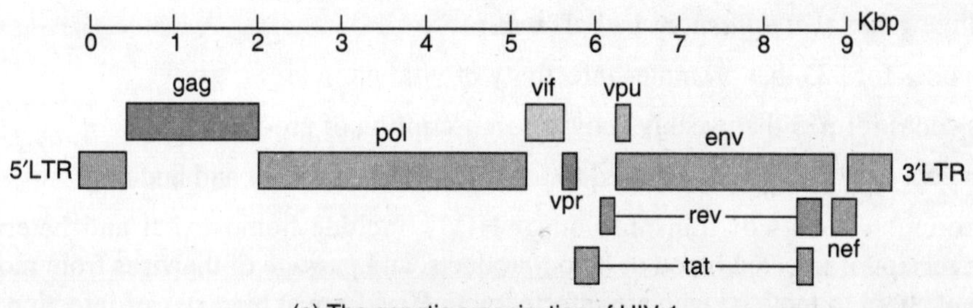

(a) The genetic organization of HIV-1

Gene	Protein product	Function of encoded proteins
gag	53-kDa precursor	*Nucleocapsid proteins*
	↓	
	p17	Forms outer core-protein layer
	p24	Forms inner core-protein layer
	p9	Is component of nucleoid core
	p7	Binds directly to genomic RNA
env	160-kDa precursor	*Envelope glycoproteins*
	↓	
	gp41	Is transmembrane protein associated with gp120 and required for fusion
	gp 120	Protrudes from envelope and binds CD4
pol	Precursor	*Enzymes*
	↓	
	p64	Has reverse transcriptase and RNase activity
	p51	Has reverse transcriptase activity
	p10	Is protease that cleaves *gag* precursor
	p32	Is integrase
vif	p23	Promotes infectivity of viral particle
vpr	p15	Weakly activates transcription of proviral DNA
tat	p14	Strongly activates transcription of proviral DNA
rev	p19	Allows export of unspliced and singly spliced mRNAs from nucleus
nef	p27	Increases viral replication; down-regulates host-cell CD4
vpu	p16	Is required for efficient viral assembly and budding

(b) Functions of the encoded proteins

Figure 13.1 Genetic organization and functions of encoded proteins of HIV-1.

There are *three regulatory genes*.

Tat gene: It encodes a protein p14 that strongly activates transcription of proviral DNA.

Nef gene: It codes for protein p27 that increases viral replication and down regulates host cell CD4.

Rev gene: It codes for protein p19 that allows export of unspliced and singly spliced mRNAs from nucleus.

There are three *genes that influence viral assembly.*

vif: This codes for p23 that promotes infectivity of viral particle.

vpr: This codes for p15 that weakly activates transcription of pro-viral DNA.

vpu: This codes for p16 that is required for efficient viral assembly and budding.

The common means of transmission of HIV-1 include homosexual and heterosexual intercourse, receipt of infected blood or blood products, and passage of the virus from mothers to infants. Infants born to mothers who are infected with HIV-1 are at high risk of infection. Unless infected mothers are treated with antiviral agents before delivery, approximately 30 per cent of infants born to them will become infected with the virus. Possible vehicles of passage from mother to infant include blood transferred in the birth process and milk in the nursing period. Transmission from an infected to an uninfected individual is most likely by transmission of the HIV infected cells, particularly macrophages, dendritic cells, and lymphocytes. While the probability of transmission by vaginal intercourse is lower than by other means, such as intravenous drug use or receptive anal intercourse, the likelihood of infection is greatly enhanced by the presence of other sexually transmitted diseases. Reasons for this increased infection rate include the lesions and organ sores present in many STDs, which favour the transfer of HIV infected blood during intercourse. At present, there is little or no evidence that casual contact with touching an infected person can spread HIV-1 infection. In well-documented cases of HIV-1 infection, there is evidence for contact with blood, milk, semen, or vaginal fluid from an infected individual. Research workers and medical professionals who take reasonable precautions have a very low incidence of AIDS, despite repeated contact with infected materials. Simple precautionary measures like use of condoms when having sex with individuals of unknown infection status is recommended.

The AIDS virus can infect human T-cells, replicating itself and in many ways causing the lysis of the host cell. The first step in HIV infection is viral attachment and entry into the target cell. HIV-1 infects T-cells that carry the CD4 antigen on their surface. The preference for $CD4^+$ cells is due to a high affinity interaction between a coat protein of HIV-1 and cell surface CD4. Expression of other cell surface molecules, co-receptors present on T-cells and monocytes, is required for HIV-1 infection. The infection of a T-cell is assisted by the T-cell co-receptor CXCR4. An analogous receptor called **CCR5** functions for the monocytes or macrophage.

After the virus has entered the cell, the RNA genome of the virus is reverse transcribed and a cDNA copy (provirus) integrates into the host genome. The integrated provirus is transcribed and the various viral RNA spliced and translated into proteins, which along with a complete new copy of the RNA genome are used to form new viral particles (Figure 13.2).

The fact that CXCR4 and CCR5 serve as co-receptors for HIV-1 on T-cells and macrophages respectively explain why some strains of HIV-1 preferentially infect T-cells (T-tropic strains) while others prefer macrophages (M-tropic strains). A T-tropic strain uses CXCR4 while the M-tropic strains use CCR5.

Screening tests for AIDS are presented in Table 13.1. The most common test is for the presence of antibodies directed against proteins of HIV-1. These generally appear in the serum of infected individuals by three months after the infection has occurred, and the individual is said to have seroconverted or to be seropositive for HIV-1. Diagnosis of AIDS include evidence for infection with HIV-1, greatly diminished numbers of $CD4^+$ T-cells (< 200 cells/μl), impaired or absent delayed hypersensitivity reactions, and occurrence of opportunistic infections (Table 13.2). Patients with AIDS generally succumb to tuberculosis, pneumonia, severe wasting diarrhoea or various malignancies. The time between acquisition of the virus and death from immunodeficiency averages 9 to 12 years. The first clear indication of AIDS may be opportunistic infection with the fungus *Candida albicans*, which causes appearance of sores in the mouth, or a vulvovaginal yeast infection in women. A persistent cough caused by *P. carinii* infection of the lungs may also be an early indicator. The virus inevitably breaks through host immune defenses, resulting in an increase in viral load, a decrease in $CD4^+$ T-cell numbers, increased opportunistic infection, and eventual death of the patient (Table 13.3).

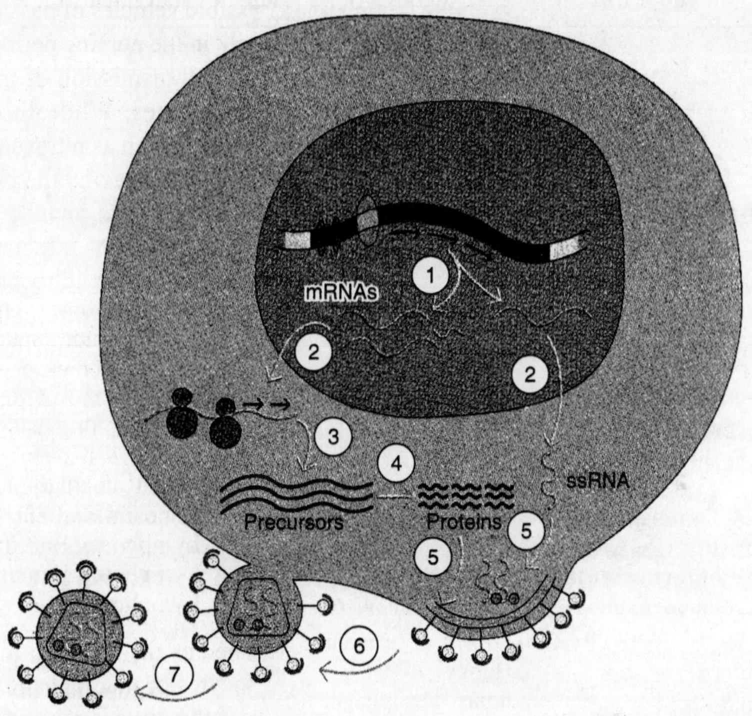

1) Transcription factors stimulate transcription of proviral DNA into genomic ssRNA and, after processing, several mRNAs.

2) Viral RNA is exported to cytoplasm.

3) Host-cell ribosomes catalyze synthesis of viral precursor proteins.

4) Viral protease cleaves precursors into viral proteins.

5) HIV ssRNA and proteins assemble beneath the host-cell membrane, into which gp41 and gp120 are inserted.

6) The membrane buds out, forming the viral envelope.

7) Released viral particles complete maturation; incorporated precursor proteins are cleaved by viral protease present in viral particles.

Figure 13.2 Activation of provirus following infection by HIV.

Table 13.1 Test Panel for AIDS

Type of Test	*Parameter Detected*	*Methods of Detection*
Screening tests	Antibodies to HIV-1 and HIV-2	EIA
Confirmatory tests	Antibodies to HIV antigens (p24, gp41, gp120/160)	Western blot test
Monitoring disease	T-cell subset ratio (helper/suppressor)	Flow cytometry (mAbs to CD3, CD4, CD8)
	HIV RNA—viral load	PCR or bDNA

Table 13.2 Clinical Diagnosis of HIV-infected Individuals

$CD4^+$ T-cell Count	*Clinical Categories*		
	A	*B*	*C*
≥ 500/μl	A1	B1	**C1**
200–499/μl	A2	B2	**C2**
< 200 μl	**A3**	**B3**	**C3**

Classification of AIDS Indicator Disease

Category A

- Asymptomatic—no symptoms at the time of HIV infection
- Acute primary infection—glandular fever-like illness lasting a few weeks at the time of infection
- Persistent Generalized Lymphadenopathy (PGL)—lymph-node enlargement persisting for 3 or more months with no evidence of infection

Category B

- Bacillary angiomatosis
- Candidiasis, oropharyngeal (thrush)
- Candidiasis, vulvovaginal: persistent, frequent, or poorly responsive to therapy
- Cervical dysplasia (moderate or severe)/cervical carcinoma in situ
- Constitutional symptoms such as fever (>38.5°C) or diarrhoea lasting >1 month
- Hairy leukoplakia, oral
- Herpes zoster (shingles) involving at least two distinct episodes or more than one dermatome
- Idiopathic thrombocytopenic purpura
- Listeriosis
- Pelvic inflammatory disease, particularly by tubo-ovarian abscess
- Peripheral neuropathy

Category C

- Candidiasis of bronchi, tracheae, or lungs
- Candidiasis, esophageal
- Cervical cancer (invasive)
- Coccidioidomycosis, disseminated or extrapulmonary
- Cryptococcosis, extrapulmonary
- Cryptosporidiosis, chronic intestinal (>1 month duration)
- Cytomegalovirus disease (other than liver, spleen, or nodes)
- Cytomegalovirus retinitis (with loss of vision)
- Encephalopathy, HIV-related
- Herpes simplex: chronic ulcer(s) (>1 month duration), bronchitis, pneumonitis, or esophagitis
- Histoplasmosis, disseminated or extrapulmonary
- Isosporiasis, chronic intestinal (>1 month duration)
- Kaposi's sarcoma
- Lymphoma, Burkitt's
- Lymphoma, immunoblastic
- Lymphoma, primary of brain
- *Mycobacterium avium* complex or *M. Kansasii*, disseminated or extrapulmonary
- *Mycobacterium tuberculosis*, any site
- *Mycobacterium*, other or unidentified species, disseminated or extrapulmonary
- *Pneumocystis carinii* pneumonia
- Progressive multifocal leukoencephalopathy
- *Salmonella* septicemia (recurrent)
- Toxoplasmosis of brain
- Wasting syndrome due to HIV

*All categories shown in bold type are considered AIDS. For category A diagnosis, no condition in categories B or C can be present; for category B diagnosis, no category C condition can be present.

Source: CDC guidelines for AIDS diagnosis.

Table 13.3 Immunologic Abnormalities Associated with HIV Infection

Stage of Infection	*Typical Abnormalities Observed*
	Lymph node structure
Early	Infection and destruction of dendritic cells; some structural disruption
Late	Extensive damage and tissue necrosis; loss of folicular dendritic cells and germinal centres; inability to trap antigens or support activation of T- and B-cells
	T helper (T_h) cells
Early	No in vitro proliferative response to specific antigen
Late	Decrease in T_h-cell numbers and corresponding helper activities; no response to T-cell mitogens or alloantigens
	Antibody production
Early	Enhanced nonspecific IgG and IgA production, but reduced IgM synthesis
Late	No proliferation of B-cells specific for HIV-1: no detectable anti-HIV antibodies in some patients
	Cytokine production
Early	Increased levels of some cytokines
Late	Shift in cytokine production from T_h1 subset to T_h2 subset
	Delayed-type hypersensitivity
Early	Highly significant reduction in proliferative capacity of T_{DTH}-cells and reduction in skin-test reactivity
Late	Elimination of DTH response; complete absence of skin-test reactivity
	T cytotoxic (T_C) cells
Early	Normal reactivity
Late	Reduction, but not elimination of CTL activity due to impaired ability to generate CTLs from T_C-cells

HIV-1 infected individuals show dysfunction of the central and peripheral nervous systems. HIV-1 infects cells in the central nervous system and viral replication occurs there. A frequent complication in later stages of HIV infection is AIDS dementia complex, a neurological syndrome characterized by abnormalities in cognition, motor performance, and behaviour.

A serological profile of HIV infection showing three stages in the infection process is depicted in Figure 13.3.

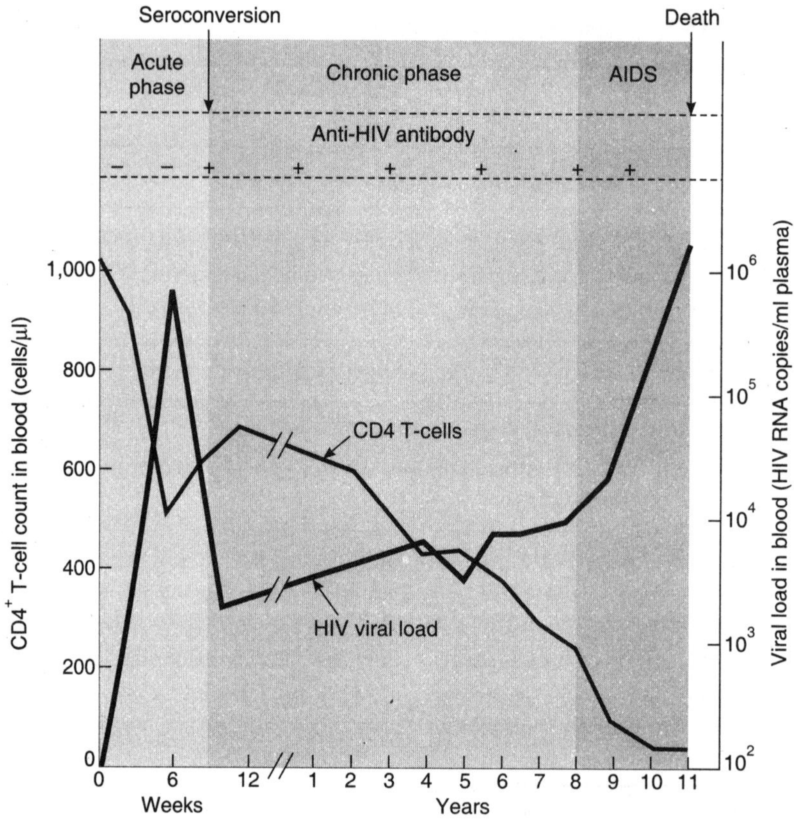

Figure 13.3 Serologic profile of HIV-infection showing three stages in the infection process.

Genetic variability of HIV-1

There is extensive variability in the genetic structure of HIV-1. The most divergent part lies within the env gene. The variable sites permit the structure of the envelope to vary in response to selective pressure from the immune response of infected individual. This permits survival of viruses that resist neutralization. The nucleotide sequence of env gene from an HIV isolate from a patient at any given time varies by 2 to 3 per cent. New variants of HIV-1 arise continuously. The reason for genetic variability is that HIV-1 reverse transcriptase copy virus RNA to proviral DNA. These enzymes are very inaccurate and make an average of one mistake per 2000 nucleotides. There is no mechanism in the virus to correct these mistakes and new variants arise within two weeks. The new HIV variants differ in their antigenicity and also their pathogenicity.

Loss of lymphocytes (lymphopenia)

The normal CD4 : CD8 count is 2. This ratio goes down to 0.5 in AIDS patients, which results in major defects in their cell-mediated immune responses. CD4 serves as a high affinity receptor for HIV-1. HIV binds to CD4 through its envelope glycoprotein gp120 and probably enters the T-cell by endocytosis. About one in 400 T-cells in blood is infected with HIV, yet only about one in 100,000 of these cells actually produce the virus. It has been shown that HIV preferentially replicates not in the blood, but in the germinal centres of the lymph nodes. Many of these germinal centre cells contain proviral DNA and not RNA, which mean that they are latently infected. Thus, significant loss of T-cells and lots of viral replication may occur in the lymph node cells while the blood T-cells appear to be relatively unaffected.

How HIV-1 kills or turns off T-cells

HIV may kill cells by any of the following ways:

(i) Infected cells simply lyse when massive amounts of virus bud off their surface.

(ii) Through cell fusion—a high level of env gene product is expressed on cells that bud viral particles. This protein causes the cells to fuse with neighbouring uninfected $CD4^+$ T-cells. The action of other cell adhesion molecules welds the cells together in a large multinuclear mass which eventually bursts. This is known as syncytium formation, which kills all the cells that had fused.

(iii) Immune destruction of T_h cells. $CD4^+$ T-cells can capture, process and present gp120 on their surface thus, making uninfected T-cells a target for HIV-1 specific cytotoxic T-cells.

Antigen may trigger apoptosis of HIV infected cells. T-cells of HIV infected individuals already carry one of two apoptosis-inducing signals. Almost any other pathogen may provide the second signal at a time which triggers apoptosis rather than an activating response.

HIV may impair CD4 T-cells by different ways which include destruction of progenitor cells; chronic partial activation leading to dysfunction; immune destruction; virus budding; syncytia formation; toxic HIV products like gp160, and accumulation of unintegrated HIV DNA. The HIV-1 subunits, like the envelope, are capable of directly inhibiting T-cell function. HIV 1 gp120 binds avidly to CD4. It may interfere with the ability of CD4 to interact with MHC Class II molecules. The ability of T-cells to respond to the presented antigen is thus, lost. The gp120 may send a negative or desensitizing signal to the T-cell thus, preventing the response to antigen. T-cells that are latently infected with HIV no longer express CD4 on their surface, are defective in synthesis and secretion of IL-2 and show decreased proliferation. There is also decreased expression of IL-2 receptor and depressed production of IFN-γ by these latently infected T-cells.

The result is a significant decline in immune function in patients whose CD4 cell numbers are normal. There is immunosuppression in AIDS patients which may be due to stimulation of cytotoxic/suppressor activity as a result of imbalance between CD4 and CD8.

B-cell abnormalities

There is an increased polyclonal activation seen in B-cells of an infected individual. Hypergammaglobulinemia, immune complex deposition and development of autoimmunity is also observed. There is a decreased response to pokeweed mitogen and new antigens, decreased memory response, decreased response to regulation and decreased isotopic switching. Epstein Barr virus transforms B-cells such that they begin expressing CD4 on their surface. HIV can infect such transformed B-cells too.

NK-cell abnormalities

The number of NK-cells is normal in AIDS patients. However their activity is suppressed. They have decreased cytotoxic activity which can be restored to normal by exposure of the NK-cells to IL-2 or to Concanavalin A. Natural killer cells are normally activated by T-cell derived IL-2.

Macrophage abnormalities

Monocytes and macrophages are reservoirs of HIV through all stages of infection. The percentage of monocytes that show CD4 is 5–90 per cent. In tissues like brain, lymph nodes and lungs, 10 to 50 per cent of macrophages are infected. HIV can replicate in macrophages, but the virions are not readily released. They accumulate in large numbers in cytoplasmic vacuoles. The macrophages then show decreased adherence to surfaces, decreased chemotaxis, decreased expression of MHC Class II molecules and decreased phagocytosis. There is also an increase in spontaneous secretion of cytokines like prostaglandins, IL-1, IL-6 and TNFα.

HIV in other cells

CD4 bearing dendritic cells that line mucosal surfaces are the first cells infected by sexually transmitted HIV. Any $CD4^+$ cell is vulnerable to HIV infection. Macrophages, monocytes, Langerhans cells, follicular dendritic cells, microglia and endothelial cells in the brain may be affected.

13.5 AIDS—THE DISEASE

Most infected people develop a fever and muscle aches three to six weeks after initial infection. This lasts for several weeks and is associated with seroconversion. Large amounts of the virus are in the blood stream and the disease can be readily transmitted. At the end of this time free virus and infected cells are eliminated immunologically. Some virus persists, and the virus replicates in low numbers for a variably long time and the patient remains fairly healthy. Eventually there is sufficient damage to the immune system, and infections and tumours begin to develop. Lymphadenopathy is characterized by development of multiple swollen lymph nodes. There is weight loss, known as HIV wasting syndrome, because of loss of appetite, altered protein metabolism, intestinal malabsorption, chronic cytokine production and endocrine abnormalities.

Development of a vaccine to prevent the spread of AIDS is on a high priority. It is critical also to develop drugs and therapies that can reverse the effects of HIV-1 infected people. The life cycle of HIV shows several susceptible points that might be blocked by pharmaceutical agents (Figure 13.4). So far, two types of antiviral agents are used. The antiviral drugs in widespread use either interfere with reverse transcription or inhibit the viral protease (Table 13.4).

Current treatments for AIDS is a combination therapy, using regimens designated HAART (Highly Active Anti-Retroviral Therapy). This combines the use of two nucleoside analogues and one protease inhibitor. This overcomes the ability of the virus to rapidly produce mutants that are drug resistant.

Table 13.4 Some Anti-HIV Drugs in Clinical Use

Generic Name (Other Names)	*Typical Dosage*	*Some Potential Side Effects*
Reverse transcriptase inhibitors: Nucleoside analog		
Didanosine (Videx, ddl)	2 pills, 2 times a day on empty stomach	Nausea, diarrhoea, pancreatic inflammation, peripheral neuropathy
Lamivudine (Epivir, 3TC)	1 pill, 2 times a day	Usually none
Stavudine (Zerit, d4T)	1 pill, 2 times a day	Peripheral neuropathy
Zalcitabine (HIVID, ddC)	1 pill, 3 times a day	Peripheral neuropathy, mouth inflammation, pancreatic inflammation
Zidovudine (Retrovir, AZT)	1 pill, 2 times a day	Nausea, headache, anemia, neutropenia (reduced levels of neutrophil white blood cells), weakness, insomnia
Pill containing lamivudine and zidovudine (Combivir)	1 pill, 2 times a day	Same as for zidovudine
Reverse transcriptase inhibitors: Nonnucleoside analogues		
Delavirdine (Rescriptor)	4 pills, 3 times a day (mixed into water); not within an hour of antacids or didanosine	Rash, headache, hepatitis
Nevirapine (Viramune)	1 pill, 2 times a day	Rash, hepatitis
Protease inhibitors		
Indinavir (Crixivan)	2 pills, 3 times a day on empty stomach or with a low-fat snack and not within 2 hours of didanosine	Kidney stones, nausea, headache, blurred vision, dizziness, rash, metallic taste in mouth, abnormal distribution of fat, elevated triglyceride and cholesterol levels, glucose intolerance
Nelfinavir (Viracept)	3 pills, 3 times a day with some food	Diarrhoea, abnormal distribution of fat, elevated triglyceride and cholesterol levels, glucose intolerance
Ritonavir (Norvir)	6 pills, 2 times a day (or 4 pills, 2 times a day if taken with saquinavir) with food and not within 2 hours of didanosine	Nausea, vomiting, diarrhoea, abdominal pain, headache, prickling sensation in skin, hepatitis, weakness, abnormal distribution of fat, elevated triglyceride and cholesterol levels, glucose intolerance
Saquinavir (Invirase, a hard-gel capsule; Fortovase, a soft-gel capsule)	6 pills, 3 times a day (or 2 pills, 2 times a day if taken with ritonavir) with a large meal	Nausea, diarrhoea, headache, abnormal distribution of fat, elevated triglyceride and cholesterol levels, glucose intolerance

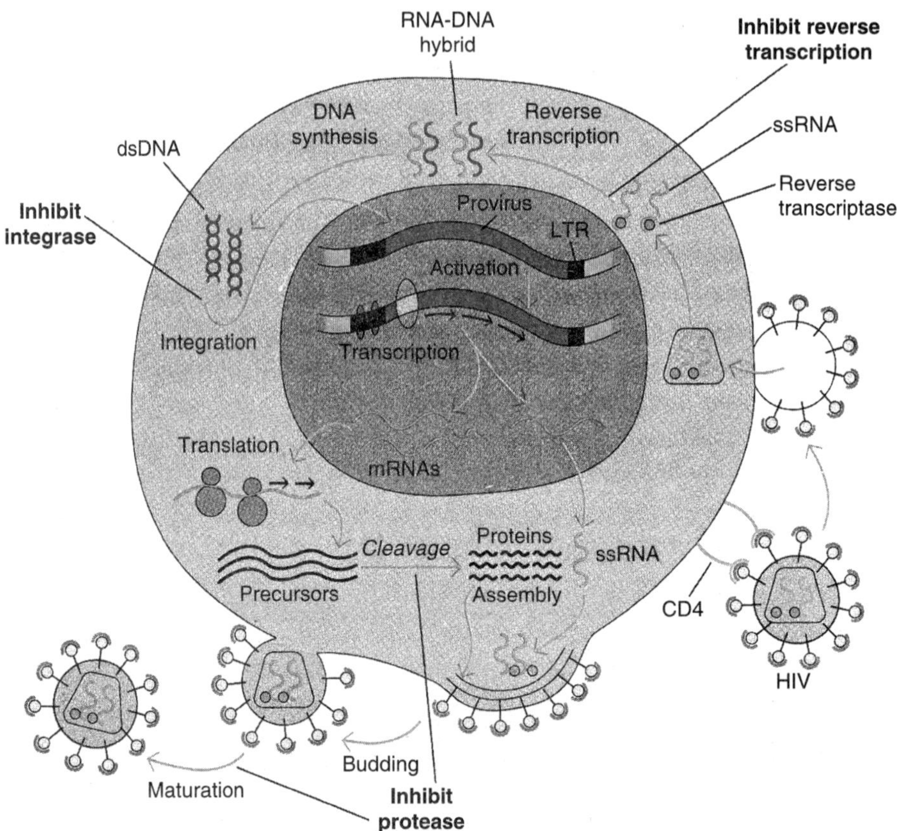

Figure 13.4 Stages in the viral replication cycle that provide targets for therapeutic antiretroviral drugs.

MULTIPLE-CHOICE QUESTIONS

1. In combined immunodeficiency there is a loss of:
   (a) B- and T-cells (b) T- and NK-cells
   (c) neutrophils and macrophages (d) eosinophils and neutrophils
2. The failure of the thymus and parathyroid to develop is called:
   (a) combined immunodeficiency (b) DiGeorge syndrome
   (c) X-linked agammaglobulinemia (d) Chediak-Higashi syndrome
3. And inherited failure to develop a functioning immune system is called:
   (a) primary immunodeficiency (b) secondary immunodeficiency
   (c) acquired immunodeficiency (d) X-linked immunodeficiency
4. Successful gene therapy has been achieved for the immunodeficiency disease:
   (a) DiGeorge anomaly (b) adenosine deaminase deficiency
   (c) diabetes mellitus (d) leucocyte adherance deficiency

5. Which component of the HIV envelop is responsible for binding to T-cells?
(a) CD4 (b) CD8
(c) gp120 (d) p24
6. The clinical deterioration seen in AIDS patients is due to a loss of:
(a) cytotoxic T-cells (b) neutrophils
(c) macrophages (d) helper T-cells
7. Which tumour is commonly observed in AIDS patients?
(a) melanoma (b) carcinoma
(c) Kaposi's sarcoma (d) Burkitt's lymphoma
8. The drug that is mainly used to treat AIDS patients is:
(a) azidothymidine (b) tetracycline
(c) cortisone (d) cyclosporin
9. It is difficult to produce a vaccine against AIDS because the HIV virus:
(a) is drug resistance (b) has a reverse transcriptase
(c) shows antigenic variation (d) hides within cells

REVIEW QUESTIONS

13.1 What are the reasons that an effective vaccine has not been developed against AIDS?
13.2 What features of HIV-1 makes AIDS an extremely difficult disease to treat?
13.3 What are the advantages and disadvantages of bone marrow transplantation?
13.4 How the deficiencies in the T-cells system differ from the deficiencies in B-cells system?

14

VACCINES

14.1 IMMUNIZATION

Immunization is the administration of an antigen to an individual to confer immunity.

Passive immunization and **active immunization** are the two methods by which an individual can be made resistant to an infectious agent (Figure 14.1). Immunity to infectious microorganisms can be achieved by any of these immunization procedures. Passive immunization produces

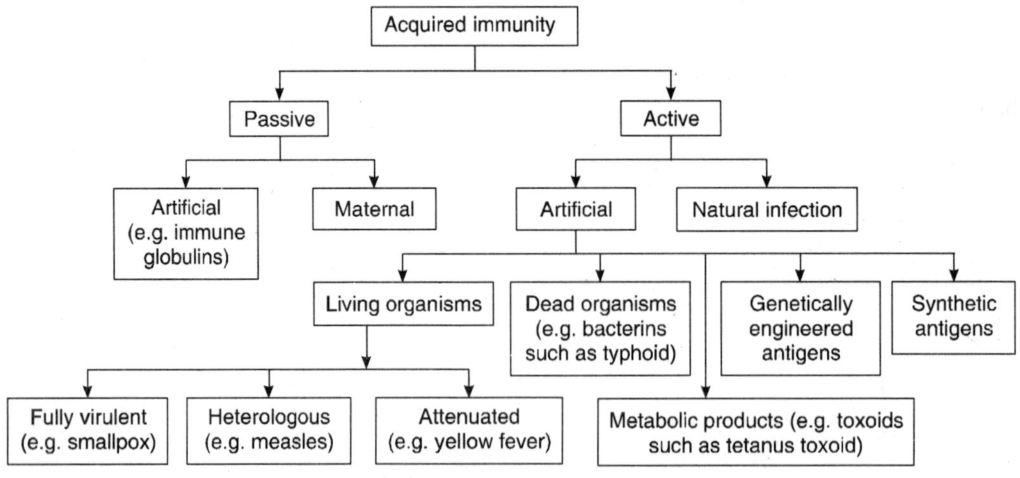

Figure 14.1 Ways in which an individual can develop immunity to an infectious agent.

temporary resistance by transferring antibodies from a resistant to a susceptible individual. These transferred antibodies give immediate protection, but since they are gradually catabolized, the protection wanes and the recipient eventually becomes susceptible to infection once again. Active immunization has several advantages over passive immunization. It involves administering antigen to individuals so that they respond by mounting a protective immune response. A later exposure to the infectious agent results in a secondary immune response. Another advantage is that it is long lasting and capable of restimulation. The major disadvantage of active immunization is that protection is not immediate.

Jenner and Pasteur are recognized as pioneers of vaccination or induction of active immunity, while Behring and Kitasato are recognized for their contributions towards passive immunity. Since then many vaccines have been developed for major afflictions of human beings. Vaccination is a cost effective process for prevention of disease.

14.1.1 Passive Immunization

In **passive immunization**, antibodies can be administered to a susceptible recipient to confer a temporary, but immediate state of immunity. This occurs naturally by transfer of maternal antibodies across the placenta to the developing foetus. Maternal antibodies to tetanus, rubella, mumps and poliovirus all afford passively acquired protection to the developing foetus. Maternal antibodies present in colostrum and milk also provide passive immunity to the infant. Passive immunization can also be achieved by injecting a recipient with preformed antibodies. Passive immunization is used under certain circumstances, like

1. If there is a deficiency in synthesis of antibody because of congenital or acquired B-cell defects, either alone or together with other immunodeficiencies.
2. When a susceptible person is exposed or likely to be exposed to a disease that will cause complications (for example, a child with leukemia exposed to varicella or measles), or when time does not permit adequate projection by active immunization.
3. When a disease is already present and the antibody may ameliorate or help to suppress the effects of toxin (tetanus, diphtheria, or botulism).

Passive immunization is usually administered to individuals exposed to tetanus, diphtheria, botulism, hepatitis, measles and rabies (Table 14.1). Passively administered antiserum is also used

Table 14.1 Common Agents Used for Passive Immunization

Disease	*Agent*
Black widow spider bite	Horse antivenin
Botulism	Horse antitoxin
Diphtheria	Horse antitoxin
Hepatitis-A and -B	Pooled human immune gamma globulin
Measles	Pooled human immune gamma globulin
Rabies	Pooled human immune gamma globulin
Snake bite	Horse antivenin
Tetanus	Pooled human immune gamma globulin or horse antitoxin

to provide protection from snakebites and black widow spider bites. Passive immunization can provide immediate protection to travellers or health-care workers who are frequently exposed to infectious organisms and lack active immunity to them. For certain diseases like acute respiratory failure in children caused by respiratory syncytial virus, passive immunization is the best treatment currently available.

In passive immunization, antibodies are produced in a donor by active immunization and these antibodies are given to susceptible animals in order to confer immediate protection. These antibodies may be raised in animals of any species and against a wide variety of pathogenic organisms. One of the most important passive immunization procedures has been the protection of humans and other animals against tetanus by means of antiserum raised in horses. The antibodies, known as **immune globulins**, are produced in young horses by a series of immunizing injections.

About 250 units of immune globulin are normally given to humans to confer immediate protection against tetanus, although the exact amount should vary with the severity of the tissue damage, the degree of wound contamination, and the time elapsed since injury. It is preferable to use antiserum of human origin whenever possible. This human Immune Serum Globulin (ISG) is obtained from volunteer donors who have recently been immunized against tetanus. Other common passive immunization agents that are used in humans include antisera against measles, hepatitis-A and -B, rabies, diphtheria and tetanus. The toxins of the clostridia are proteins, and they can be made non-toxic by treatment with formaldehyde. The toxins treated in this way are known as **toxoids**.

Monoclonal antibodies represent another potential source of passive protection. At present, these are mainly made by mouse-mouse hybridomas and thus consists of mouse immunoglobulins. They are therefore, able to sensitize animals of other species.

Passive immunization also has some adverse consequences. Since passive immunization does not activate the immune system, it generates no memory response and the protection provided is transient. The immunity achieved is short-lived and there is suppression of active immunization. It sometimes results in serum sickness (if Ig is still circulating when recipient begins to form antibodies) or in anaphylaxis. If repeated doses of horse immune globulin are given to an individual it gives rise to hypersensitivity.

There are also certain risks involved in passive immunization. If the antibody was produced in another species, such as the horse, the recipient can mount a strong response to the isotypic determinants of the foreign molecule. Some individuals produce IgE antibody specific for a passive antibody, which can mediate systemic mast cell degranulation, leading to systemic anaphylaxis. Thus the aim of passive immunization is transient protection or alleviation of an existing condition.

14.1.2 Active Immunization

Immunization procedures involve giving antigen derived from an infectious agent to an individual so that an immune response is mounted and resistance to that infectious agent is stimulated. This is known as **active immunization**.

The goal of active immunization is to elicit protective immunity and immunologic memory.

Active immunization through vaccination plays a very important and active role in proliferation of antigen sensitive T- and B-lymphocytes leading to the formation of immune memory cells. To achieve lifelong immunity, childhood immunization requires multiple boosters at appropriate time intervals. For instance, multiple immunizations with oral polio vaccine is necessary to ensure a sufficient immune response to be generated against each of the three strains of polio virus that make up the Sabin virus.

Advantages of active immunization are the prolonged duration of protection and the recall and boosting of this protective response by repeated injections of antigen. Active immunization can be achieved by natural infection with a microorganism, or it can be acquired artificially by administration of a vaccine (Table 14.2). In active immunization, proliferation of antigen-reactive T- and B-cells results in the formation of memory cells.

Table 14.2 Acquisition of Passive and Active Immunity

Type	*Acquired Through*
Passive immunity	Natural maternal antibody
	Immune globulin
	Antitoxin
Active immunity	Natural infection
	Vaccines
	–Attenuated organisms
	–Inactivated organisms
	–Purified microbial macromolecules
	–Cloned microbial antigens (alone or in vectors)
	–Multivalent complexes
	Toxoid

14.2 VACCINES

A preparation of antigenic material used to induce immunity against pathogenic organisms is a vaccine. It can be a suspension of attenuated live or killed microorganisms, or antigenic portions of them, presented to a potential host to induce immunity and prevent disease.

For developing a successful vaccine, several factors must be kept in mind. First, the development of an immune response does not necessarily mean that a state of protective immunity has been achieved. A second factor is the development of immunologic memory. If living organisms must be used in a vaccine, they must be treated in such a way that they lose their disease producing ability. This is known as **attenuation**, which involves subjecting adaptive organisms to unusual environmental conditions so that they lose their ability to replicate uncontrollably in their usual host. For example, the Bacillus Calmette-Guerin (BCG) strain of *Mycobacterium bovis* was rendered avirulent by growing it for 13 years on a bile-saturated medium. At the end of that time, the organisms had adapted well to growing in bile, but they had lost the ability to grow and cause disease in humans.

Viruses also can be attenuated by growth in abnormal culture conditions. Prolonged tissue culture, especially in cells of the species that the virus does not normally infect, has the effect of reducing the virulence of viruses. Another method of attenuation is to adapt the virus to growth at slightly lower temperatures than normal. It then cannot spread within the body and so cause disease, because the body temperature is too high.

Attenuated vaccines: Pathogenicity of certain microorganisms can be attenuated so that they lose their ability to cause disease. Attenuated organisms however retain their capacity to undergo transient growth in an inoculated host. Attenuation of a bacteria or virus is achieved by growing them for a prolonged period under abnormal culture conditions, like in an unnatural host. Those mutants which are better suited to growth in the abnormal culture conditions; but fail to grow to such an extent in the original host, are selected. Sabin vaccine for polio was obtained by growing polio virus in monkey kidney epithelial cells. The measles vaccine contains a strain of rubella virus which was attenuated by growing in duck embryo cells and later in human diploid cell lines. Attenuated vaccines have a number of advantages. The ability of many attenuated vaccines to replicate within the most cells make them particularly suitable for inducing cell-mediated immune response. Attenuated vaccines show transient growth, thereby providing prolonged immune system exposure to individual epitopes on the attenuated organisms. This results in increased immunogenicity and memory cell production. Hence, such vaccines often require only a single dose of administration. The disadvantages of attenuated vaccines include the possibility of their reversion into virulent forms. The rate of reversion in Sabin polio vaccine is one in 4 million doses of vaccine. Sometimes post vaccination complications may develop in individuals.

The type of antigen used in a vaccine depends on many factors. The more antigens of the microbe retained in the vaccine, the better. Moreover, living organisms tend to be more effective than killed ones Table 14.3.

Table 14.3 Antigenic Preparations that are Used as Vaccines

	Type of Antigen	*Vaccine Examples*
Living organisms	Natural	Vaccinia (for smallpox) Vole bacillus (for TB; historical)
	Attenuated	Polio (Sabin; oral polio vaccine) Measles, mumps, rubella Yellow fever 17D Varicella-zoster (human herpes virus 3) BCG (for TB)
Intact, but non-living organisms	Viruses	Polio (Salk), rabies, influenza, hapatitis-A, typhus
	Bacteria	Pertussis, typhoid, cholera, plague
Subcellular fragments	Capsular polysaccharides	Pneumococcus, Meningococcus, *Haemophilus influenzae*
	Surface antigen	Hepatitis B
Toxoids		Tetanus, diphtheria
Recombinant DNA-based	Gene cloned and expressed	Hepatitis-B (yeast-derived)
	Genes expressed in vectors	Experimental
	Naked DNA	Experimental
Anti-idiotype		Experimental

Natural live viruses have rarely been used because of safety problems. Attenuated live viruses have been highly successful. The preferred strategy is to attenuate a human pathogen, with the aim of diminishing its virulence while retaining the desired antigens. Attenuation can also be achieved by mutation. The Type I polio vaccine contains 57 mutations, and has almost never reverted to the wild type. With the recombinant DNA technology now available, it seems likely that the future attenuated viral and bacterial vaccines will have site-directed rather than random mutations.

Vaccines containing living organisms are effective and give prolonged, strong immunity. They cause transient infection, so only a few inoculating doses are needed and no adjuvant is required. They may also provoke a rapid protective response through stimulation of interferon production. However, live vaccines are difficult and expensive to produce. They may cause disease because of some residual virulence or reversion to a more virulent form. Moreover they cannot be used in patients suffering from an immunodeficiency.

Killed or inactivated organisms are also used for making vaccines. The pathogen is inactivated by heat or by chemicals; hence, it is no longer capable of replicating in the host. Heating kills, but causes extensive protein denaturation, and antigenic similarity may be lost. Through chemical inactivation, chemicals must produce little change in the antigens that are responsible for protective immunity. It is important to maintain the structure of epitopes on the surface antigens during the process of inactivation. Vaccines against polio and whooping cough are produced by formaldehyde treatment of the pathogen. Killed vaccines predominantly induce humoral immune response, whereas attenuated vaccines induce cell mediated as well as humoral immune response. Instances where quality control procedures are not properly followed while producing killed vaccines may lead to complications in the host.

The inactivating agents that are normally used are formaldehyde, ethylene oxide, acetyl ethylene amine, ethylene amine, and β-propiolactone.

Vaccination involves adaptive immunity. The art of vaccination is to produce antigenic preparations from the pathogens that are safe to administer, induce the right sort of immunity and are affordable by the population at which they are aimed.

14.3 DESIGNING VACCINES

The purpose of designing vaccines is to achieve a state of protective immunity and the development of immunologic memory.

An ideal vaccine to be used for active immunization should conform to certain parameters. It should be

1. Cheap
2. Stable
3. Adaptable to mass vaccination
4. Able to give prolonged strong immunity
5. Have no adverse side-effects
6. Stimulate an immune response distinguishable from that due to natural infection, so that vaccination does not interfere with diagnosis.

14.4 ROUTE OF ADMINISTRATION

Injectable vaccines should be administered in an area that has the least risk of tissue injury. Injections are relatively time-consuming and painful and bear a risk of carrying unwanted organisms into an individual. When large numbers of people must be vaccinated under less than ideal conditions, then other methods of vaccination are employed. A high-pressure jet injector serves as a painless and sterile method of administering antigen. Vaccines against polio or typhoid are given orally. In the poultry industry vaccines are administered to large flocks by incorporating the vaccine in the feed or in the drinking water. Animals can be exposed to a vaccine aerosol so that they will inhale the vaccine.

In vaccination, it is sometimes desirable to administer an **adjuvant** along with the antigen to enhance the normal immune responses.

In short, vaccines may be administered in any of the following ways:

1. Intramuscular or subcutaneous injections
2. High-pressure jet injection which is painless and sterile
3. Oral route (polio and typhoid vaccines in humans, and vaccine given in the feed or drinking water to poultry)
4. Inhalation (vaccine aerosol used in veterinary medicine). This is used particularly for vaccines directed against diseases of the respiratory tract
5. Along with an adjuvant, e.g. vaccines used against diphtheria, whooping cough and tetanus.

To determine whether vaccination is either possible or desirable in controlling a specific disease, three criteria must be satisfied. The first is the absolute identification of the causal organism. Second, it must be established that an immune response can protect against the disease in question. Finally, one must be absolutely certain that the risks involved do not exceed those, due to the chance of contracting the disease itself.

14.5 TYPES OF VACCINES

14.5.1 Anti-idiotype Vaccines

Antigens provoke the appearance of immunoglobulin molecules whose binding site (idiotype) has a complimentary structure to that of the inducing epitope. Antibodies to this idiotype have the same three-dimensional structure as the inducing epitope. Anti-idiotype antibodies can be made by taking a monoclonal antibody against the epitope in question and using it to immunize an animal. The anti-anti-idiotypes formed may be protective, directed against not only the anti-idiotype, but also the original antigen.

14.5.2 Whole Organism Vaccines

Vaccines containing living organisms

These are effective and give prolonged, strong immunity. Only a few inoculating doses are required and adjuvant need not be employed, lowering the chances of hypersensitivity reactions. Moreover, live virus vaccines may provoke a rapid protective response through stimulation of interferon production. On other hand, live virus vaccines may be difficult and expensive to produce. They may contain dangerous extraneous organisms, and may cause disease as a result either of residual virulence or of reversion to a more virulent form. Live vaccines cannot be used on patients suffering from an immunodeficiency.

To avoid these disadvantages, it is common to use organisms that have been killed or inactivated. The dead organisms should be as antigenically similar to the live organisms as possible. Therefore, heat killing, which causes extensive protein denaturation, is an unsatisfactory procedure. Chemical inactivation is thus, done using inactivating agents like formaldehyde, ethylene oxide, ethyleneimine, acetyl-ethyleneimine and β-propiolactone.

New approaches are being studied in attempts to make vaccines more effective, cheaper and safer (Figure 14.2).

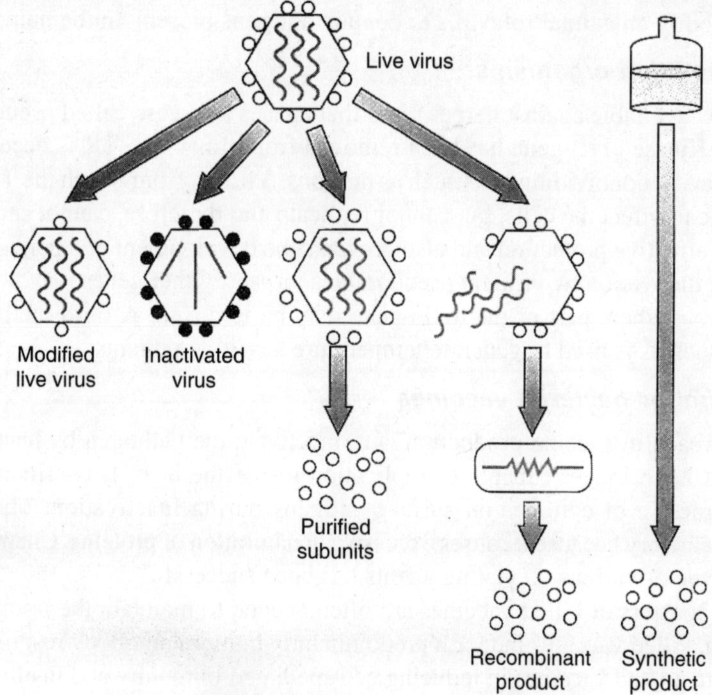

Figure 14.2 Some methods of making vaccines.

Attenuated viral or bacterial vaccines

Microorganisms can be attenuated so that they loose their ability to cause significant disease, but retain their capacity for transient growth within an inoculated host. Attenuation can be achieved by growing a pathogenic bacterium or virus for prolonged periods under abnormal culture conditions. This procedure selects mutants that are better suited to growth in the abnormal culture conditions and are therefore, less capable of growth in natural host. An example earlier cited is BCG vaccine which was developed after the strain had adapted to growth with increased bile in thirteen years and had become sufficiently attenuated so that it was suitable as a vaccine for tuberculosis.

Attenuated vaccines have some advantages and some disadvantages. Because of their capacity for transient growth, such vaccines provide prolonged exposure of the individual epitopes on the attenuated organisms, to the immune system resulting in increased immunogenicity and production of memory cells. Thus, these vaccines often require only a single immunization, eliminating the need for repeated boosters. A major disadvantage of attenuated vaccines is the possibility of their reversion to a virulent form. Another concern with attenuated vaccines is the presence of other viruses as contaminants. In some cases most vaccine complications may render a potential vaccine unacceptable for use.

Genetic engineering techniques provide a way to attenuate a virus irreversibly by selectively removing genes that are necessary for virulence. Genes of organisms can be deliberately modified so that they become effectively and irreversibly attenuated. It is possible that genetic engineering techniques could eliminate the risk of reversion of attenuated polio vaccine. More recently, a vaccine against rotavirus, a major cause of infant diarrhoea, was developed using genetic engineering techniques to modify an animal rotavirus to contain antigens present on the human viruses.

Genetically modified organisms

A vaccine is now available against herpes virus that causes a disease called pseudorabies in pigs. The Thymidine Kinase (TK) gene has been removed from this virus. TK is necessary for herpes viruses to replicate in nondividing cells such as neurons. Viruses from which the TK gene has been removed are able to infect the cells, but cannot replicate and therefore, cannot cause disease. This vaccine confers effective protection and also prevents the development of a persistent carrier state by blocking cell innovation by virulent pseudorabies viruses. Other genes may be deleted such as in *Vibrio cholerae*, where part of the toxin gene may be removed. A third method is to insert a missense gene such as is used to generate temperature sensitive mutants.

Inactivated viral or bacterial vaccines

A common approach in vaccine production is to inactivate the pathogen by heat or by chemical means so that it is no longer capable of replication inside the host. It is critically important to maintain the structure of epitopes on surface antigens during inactivation. The inactivation is generally unsatisfactory because it causes excessive denaturation of proteins. Chemical inactivation with formaldehyde or various alkylating agents has been successful.

Repeated boosters of killed vaccines are often needed to maintain the immune status of the host. In addition, killed vaccines induce a predominantly humoral antibody response; they are less effective than attenuated vaccines in inducing cell-mediated immunity and in eliciting a secretory IgA response.

14.5.3 Recombinant Vaccine

Recombinant DNA techniques can be employed to isolate genetic material coding for protein antigen of interest. For making such vaccines the genes that encode major antigens of virulent pathogens are introduced into attenuated viruses or bacteria. The attenuated organism serves as a vector, replicating within the host and expressing the gene product of the pathogen. These techniques are useful in any situation in which protein antigen is to be synthesized in large and pure quantities. But pure proteins are often poor antigens because they are not effectively delivered to antigen sensitive cells and therefore, cannot provoke a sufficiently intense response.

Organisms that have been used as vectors for vaccines are vaccinia virus, attenuated polio virus, adenovirus, BCG strain of Mycobacterium Bovis and the attenuated strain of salmonella.

The degree of expression of inserted foreign pathogenic gene is quite high in vaccinia virus, which serves as a potent immunogen and inoculated host. The gene for hepatitis-B surface antigen, and influenza have been successfully inserted into vaccinia genome. Vaccines produced from other activated vectors may prove safer than the vaccinia vaccines, because vaccinia may become virulent in individuals suffering from severe immune suppression.

Vaccinia virus is most widely employed vector as it is easy to administer by dermal scratching. It is a large complex virus with the genome of about 200 genes; it can be engineered to carry several dozen foreign genes without impairing its capacity to infect host cells and replicate. It expresses high levels of the new antigen. Moreover, the proteins undergo appropriate processing steps, including glycosylation and membrane transport.

A number of genes encoding surface antigens from viral, bacterial, and protozoan pathogens have been successfully cloned into bacterial, yeast, insect, or mammalian expression systems, and the expressed antigens used for vaccine development. The first such recombinant antigen vaccine approved for human use is the hepatitis-B vaccine.

DNA vaccines

Plasmid DNA encoding antigenic proteins is injected directly into the muscle of the recipient. The muscle cells take up the DNA and the encoded protein antigen is expressed, leading to both a humoral antibody response and the cell-mediated response. The injected DNA is taken up and expressed by muscle cells with much greater efficiency than in tissue culture. Dendritic cells in the area that take up the plasmid DNA and express the viral DNA also express the viral antigen.

It is not exactly known whether the injected DNA integrates with the chromosomal DNA of muscle cells or remains as episomal body. DNA vaccines have many benefits. They are inexpensive to produce, easily manipulated, noninfectious, can induce both humoral and cellular responses, and they only express the protein of interest.

DNA vaccines offer advantages over many of the existing vaccines. The encoded protein is expressed in the host in its natural form and there is no denaturation or modification. The immune response will be directed to the antigen exactly as expressed by the pathogen. They also cause prolonged expression of the antigen, which generates immunological memory. Refrigeration is not required for the handling and storage of the plasmid DNA. The same plasmid vector can be customized to make a variety of proteins, so that the same techniques can be used for different DNA vaccines, each encoding an antigen from a different pathogen.

Human trials are underway for vaccines for malaria, herpes virus, AIDS and influenza.

A drawback is that only protein antigens can be encoded. Another shortcoming is the inability to use DNA immunization with vaccines such as oral vaccines or those given as a nasal spray.

Naked DNA

Philip Felgner and Robert Malone developed a unique approach to vaccination that involves injection of DNA from a virus into a host. The gene products are treated as endogenous antigens, processed by linkage to MHC Class I molecules, and displayed on the cell surface. This processed antigen then triggers a protective immune response. This technique can be refined even further since the viral DNA can be shot directly into cells when coated onto tiny gold beads fired by a *gene gun*.

14.5.4 Synthetic Peptide Vaccines

Globular protein molecules have a limited number of epitopes on the surface. Only a few of these epitopes are important in inducing protective immunity. If the structure of a protective epitope is known, it may be chemically synthesized and used in a vaccine. The use of synthetic peptides as vaccines has not progressed quickly. Peptides are not as immunogenic as proteins and it is difficult to elicit both humoral and cellular immunity to them. The construction of synthetic peptides for use as vaccines to induce either a humoral or cell-mediated immunity requires an understanding of the nature of T-cell and B-cell epitopes. Vaccines for inducing humoral immunity should include peptides that compose immunodominant B-cell epitopes. An effective memory response for both humoral and cell-mediated immunity requires generation of a population of memory T_h-cells. A successful vaccine must therefore, include immunodominant T-cell epitopes too. For developing a synthetic peptide vaccine, helper and suppressor peptides within an antigen are identified. By eliminating suppressor peptides from synthetic vaccines we can have enhanced immunity. For the preparation of synthetic peptide vaccines the nature of B-cell and T-cell epitopes must be understood. Synthetic peptides are designed so that they strongly bind to hydrophilic region of the B-cell epitope. A successful vaccine must also have an immunodominant T-cell epitopes.

Experimental synthetic vaccines have been developed against several viruses including hepatitis-B and influenza-A. This technique is much safer than genes splicing techniques, but it is considerably more expensive.

Synthetic peptide vaccines for HIV, influenza, diphtheria toxin, hepatitis-B virus and malaria parasite are currently being evaluated for their efficacy.

The disadvantage is that there is less cell-mediated immune response.

14.5.5 Multivalent Subunit Vaccines

Synthetic peptide vaccines and recombinant protein vaccines tend to be poorly immunogenic. They induce a humoral antibody response, but do not induce a good cell mediated response. Synthetic peptide vaccines should contain both immunodominant B-cell and T-cell epitopes. One approach is to prepare Solid Matrix Antibody Antigen (SMAA) complexes by attaching monoclonal antibodies to particulate solid matrices and then saturating the antibody with the desired antigen. The resulting complexes are then used as vaccines. By attaching different monoclonal antibodies to the solid matrix, it is possible to bind a mixture of peptides or proteins, composing immunodominant epitopes for both T-cells and B-cells, to the solid matrix (Figure 14.3). These complexes induce vigorous humoral and cell mediated responses. Their particulate nature contributes to their increased immunogenicity by facilitating phagocytosis. Multivalent vaccines are also produced by using detergent to incorporate protein antigens or synthetic antigenic peptides into protein micelles, into lipid vesicles called **liposomes**, or into immunostimulating complexes. ImmunoStimulating Complexes (ISCOMs) are lipid carriers prepared by mixing protein or peptide antigens with the detergent and a glycoside called **Quil A**.

Potential vaccines are membrane proteins from influenza virus, hepatitis-B virus, HIV incorporated into micelles, liposomes and ISCOMs.

Liposomes and ISCOMs fuse with the plasma membrane and deliver the antigen in intracellular area where it can be processed through cytosolic pathway and induce cell mediated immune response. Presently protein components of influenza virus, measles virus, HIV and hepatitis-B are incorporated into ISCOMs, liposomes and micelles, and are being assessed for their vaccine properties.

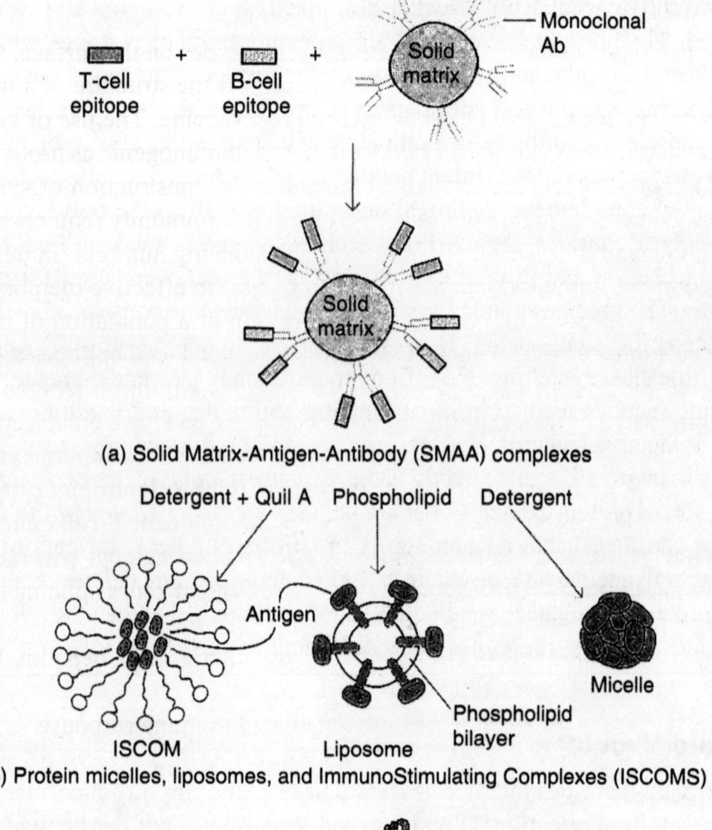

(a) Solid Matrix-Antigen-Antibody (SMAA) complexes

(b) Protein micelles, liposomes, and ImmunoStimulating Complexes (ISCOMS)

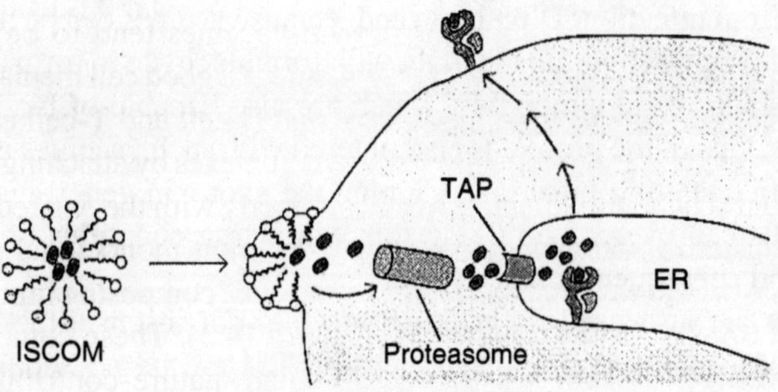

(c) ISCOM delivery of antigen into cell

Figure 14.3 Multivalent subunit vaccines.

14.5.6 Purified Macromolecules as Vaccines

Vaccines that consist of specific, purified macromolecules derived from pathogens do not have the risks associated with attenuated or killed whole organism vaccines.

Polysaccharide vaccines

These are vaccines from purified macromolecules. The risks associated with attenuated or killed whole organism vaccines can be overcome by preparing vaccines from purified macromolecules of pathogen. Currently capsular polysaccharides, inactivated exotoxins and recombinant surface antigens are used. The polysaccharides contain repeating units of epitopes within one molecule which result in cross-linking of antigen receptors on the surface of the lymphocytes. Polysaccharide capsules of certain bacteria possess antiphagocytic properties and therefore, they show virulence. Coating of the capsule by antibody or complement increases the ability of the neutrophils and macrophages to phagocytose such virulent pathogens. Many bacterial polysaccharides can activate alternate pathway of complement leading to deposition of C3b on the polysaccharides.

Capsular polysaccharides are used as vaccines against influenza virus where the type of vaccine is polysaccharide and protein carrier. Polysaccharides were used for making a vaccine against meningitis. The vaccine against *Streptococcus pneumoniae* consists of 23 different antigenic capsular polysaccharide components. However, there is no activation of T_h-cells. There is IgM production, but little class switching. No affinity maturation is seen and there are very less memory cells. The vaccine induces formation of opsonizing antibodies and is administered to high-risk groups such as infants, splenectomized patients, other immune suppressed individuals, and the elderly. One way to involve T_h-cells directly in the response to polysaccharide antigen is to conjugate antigen to some sort of protein carrier. For example, the vaccine for *Haemophilus influenzae* Type b (Hib), the major cause of bacterial meningitis in children under five years of age, consists of Type b capsular polysaccharide covalently linked to a protein carrier, tetanus toxoid. The polysaccharide protein conjugate is considerably more immunogenic than the polysaccharide alone, because it activates T_h-cells, it enables class switching from IgM to IgG and it can also induce memory B-cells.

14.5.7 Toxoid Vaccines

Some bacterial pathogens produce exotoxins. These exotoxins produce many of the disease symptoms that result from infection. Diphtheria and tetanus vaccines can be made by purifying the bacterial exotoxin and then inactivating the toxin with formaldehyde to form a toxoid. Vaccination with a toxoid induces antitoxoid antibodies, which are also capable of binding to the toxin and neutralizing its effects. One of the problems encountered with such vaccines is obtaining sufficient quantities of the purified starting materials. Cloning the exotoxin genes and expressing them in easily grown host cells can overcome this. In this way, large quantities of the exotoxin can be produced, purified, and subsequently inactivated.

Surface antigens are also used for making vaccine. For instance the recombinant surface antigen (HbsAg) is used to make vaccine against hepatitis-B. Here the gene coding any immunogenic protein can be cloned and expressed in bacterial, yeast or mammalian cells. The expressed antigens are used for vaccine development.

There are three distinct roles of heat shock proteins which are involved in the immunogenic or immunogenicity modulated effects. (i) Microbial Hsps are immunodominant antigens (ii) Hsps as adjuvants in vaccines (iii) Hsps as chaperones in vaccines that facilitate T-cell priming.

14.6 VACCINATION SCHEDULE

A simple immunization schedule can be followed as children grow up (Figure 14.4). It may be modified as appropriate if vaccination had begun at a later age than normal. Maternal antibodies passively protect newborn animals and therefore, it is sometimes difficult to successfully vaccinate them in early life. In domestic animals, if protection is required early in life, it may be possible to vaccinate the pregnant mother so that peak antibody levels are achieved at the time of colostrum formation. The interval between booster doses of vaccine varies, but inactivated vaccines generally produce a weak immunity that requires frequent boosters. Live vaccines produce a much more persistent immunity. For example, BCG vaccine does not require boosting, and yellow fever vaccine requires a booster dose only every ten years (Tables 14.4 and 14.5).

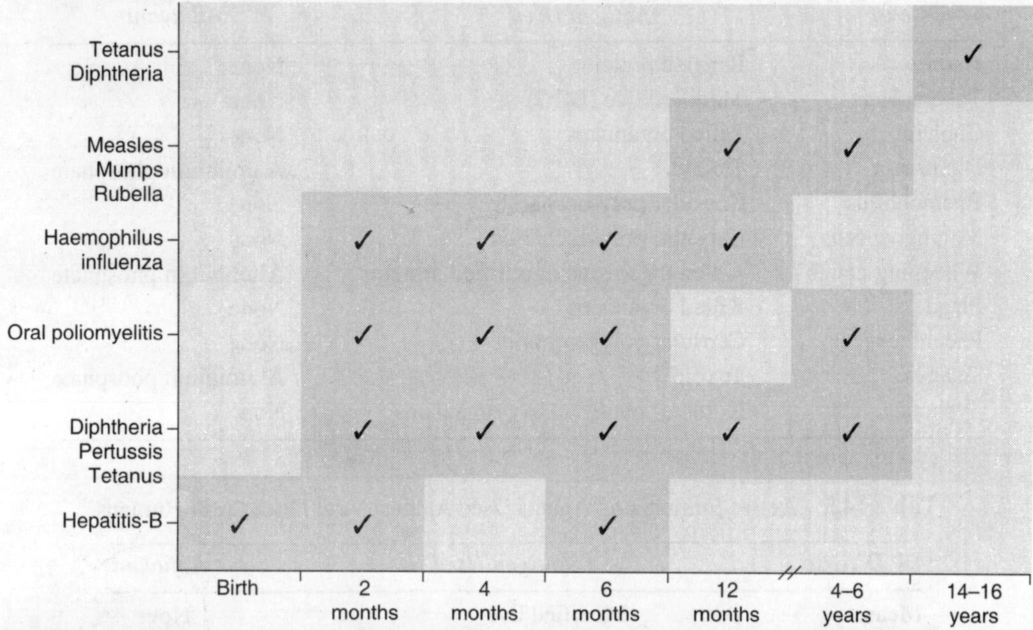

Figure 14.4 Routine immunization schedule for normal infants and children.

In the adults, vaccination varies depending on the severity of the disease. Vaccination for meningitis, pneumonia and influenza is given to individuals living in larger groups like military camps and excursion groups. International travellers are also vaccinated against a number of diseases like cholera, typhoid, yellow fever, plague, etc.

14.7 FAILURES IN VACCINATION

The immune response never confers absolute protection and it is never equal in all members of a vaccinated population. The immune response is influenced by many factors. Although most individuals respond to a vaccine by mounting an average immune response, a small proportion mounts a poor immune response. This group may not be protected against the disease inspite of vaccination. The size of the unreactive portion of the population varies with the type of vaccine employed and the number of booster doses given. From a public health viewpoint, less than hundred per cent protection may be quite satisfactory if it prevents the spread of disease within the population. This phenomenon is known as **herd immunity**; its effect is due to the reduced probability of a susceptible individual encountering an infected one and so contracting disease.

A second type of apparent vaccine failure can occur when the immune response is depressed. This may be due to clinical immunosuppression or due to a high stress situation. Extremes of heat and cold or malnourishment may inhibit the immune response. Other reasons include the possibility that the individual was incubating the disease prior to vaccination, and also the excessive use of alcohol while swabbing the skin, which may inactivate a virus vaccine.

Table 14.4 Active Immunizing Agents Used Against Bacterial Diseases in Humans

Disease	*Antigen Used*	*Adjuvant*
Anthrax	Purified proteins	None
Tuberculosis	Modified live (BCG)	None
Cholera	Killed organisms	None
Diphtheria	Toxoid	Aluminium phosphate
Haemophilus	Capsular polysaccharide	None
Meningococcus	Capsular polysaccharide	None
Whooping cough	Killed organisms or purified proteins	Aluminium phosphate
Plague	Killed organisms	None
Pneumococcus	Capsular polysaccharide	None
Tetanus	Toxoid	Aluminium phosphate
Typhoid	Killed or modified live organisms	None

Table 14.5 Active Immunizing Agents Used Against Viral Diseases in Humans

Disease	*Antigen Used*	*Adjuvant*
Measles	Modified live	None
Mumps	Modified live	None
Rubella	Modified live	None
Poliomyelitis	Inactivated or modified live	None
Influenza	Inactivated	None
Hepatitis-B	Recombinant antigen	None
Rabies	Inactivated	None
Yellow fever	Modified live	None

MULTIPLE-CHOICE QUESTIONS

1. The major disadvantage of passive immunization is that it:
   (a) is ineffective
   (b) induces short lived immunity
   (c) interferes with active immunization
   (d) is expensive

2. The major disadvantage of active immunization that it:
   (a) causes prolonged immunity
   (b) induces rapid onset of immunity
   (c) induces slow onset of immunity
   (d) is expensive
3. Organisms suitable for use in modified live vaccines are produced by:
   (a) inactivation (b) genetic recombination
   (c) attenuation (d) hybridization
4. A suitable organism for use in recombinant vaccines is:
   (a) influenza virus (b) smallpox virus
   (c) poliovirus (d) vaccinia virus
5. A gene that is involved in one mutation that makes herpes virus avirulent is:
   (a) envelop gene (b) thymidine kinase gene
   (c) nucleocapsid gene (d) polymerase gene
6. The most potent adjuvant known is:
   (a) alum (b) endotoxin
   (c) Freund's adjuvant (d) salt solution
7. Which of the following makes vaccines fail in very young infants?
   (a) maternal antibodies (b) glycoproteins
   (c) serum (d) endotoxins
8. A problem with the use of modified live vaccine is:
   (a) toxicity (b) residual virulence
   (c) encephalitis (d) muscle pain

REVIEW QUESTIONS

14.1 How is immune globulin manufactured against snake venom?
14.2 What are vaccines?
14.3 What is the role of heat shock proteins in DNA vaccines?
14.4 What are the different types of vaccines?
14.5 What things should be kept in mind while designing vaccines?

15

CYTOKINES

15.1 CYTOKINES

The word cytokine is derived from two words: 'Cyto' which means 'cell' and 'kinein' which means 'to move'. Cytokines have an important role in cell-to-cell communication. **Cytokines** are a group of low molecular weight regulatory proteins/glycoproteins. They regulate the intensity and duration of immune response by

1. Stimulating or inhibiting the activation, proliferation and differentiation of cells.
2. Regulation of secretion of antibodies or other cytokines.

Cytokines are secreted by white blood cells and a variety of other cells in the body in response to various types of stimuli. They regulate the development of immune effector cells and may have direct effector functions. They bind to specific receptors on membranes of the target cells to show their effect.

Cytokines are named according to their origin and function. Thus, we have:

Lymphokines: Cytokines secreted by lymphocytes.

Monokines: Cytokines secreted by monocytes and macrophages.

Interleukins: Secreted by leukocytes and act on other leukocytes affecting their growth and differentiation.

Chemokines: Low molecular weight cytokines that effect chemotaxis for different leukocytes and regulate the expression and adhesiveness of leukocyte integrins.

Cytokines generally have a molecular weight less than 30 kDa. Structural studies show that there are four families of cytokines.

1. Haematopoeitin family
2. Interferon family
3. Chemokine family
4. Tumour necrosis factor family

Cytokines bind to specific receptor on the target cell membrane and induce signal transduction that ultimately results in triggering gene expression in target cells.

The binding of cytokines to responsive target cells generally stimulates increased expression of cytokine receptors or increased secretion of other cytokines, which affect other target cells in turn. For example, cytokines produced by T_h-cells influence the activities of B-cells, T_C-cells, NK-cells, macrophages, granulocytes and haematopoietic stem cells. Thus, an entire network of cells is mobilized.

15.2 CYTOKINE RECEPTORS

Cytokine receptors are made up from several different chains. Receptor for a specific cytokine may exist in various combinations of these component chains. These component chains may vary in the affinity in which they bind the cytokine, or in the ability to initiate a signal transducing pathway after binding.

15.2.1 Families of Cytokine Receptors

There are five families of cytokine receptors.

1. Immunoglobulin superfamily receptors: Receptors for IL-1 are of two types.Type1 IL-1R is expressed on many cell types, while Type 2 IL-1R is limited to B-cells.

2. Class I cytokine receptor family (haematopoietin receptor family): Majority of cytokine-binding receptors that function in immune and haematopoietic system are included in this class. They have conserved amino acid sequence motifs in the extra cellular domain, (CCCC) and (WSXWS).

3. Class II cytokine receptor family (interferon receptor family): These possess CCCC motif, but not the WSXWS. They bind α, β, γ interferons, six members of IL-10 family, 17 Type I interferons, 1 Type II interferon, three members of λ IFN, IL-28a, IL-28-b and IL-29.

Both Class I and II cytokine receptor families have multiple subunits including one that binds specific cytokine molecules and another that mediates signal transduction.

4. TNF receptor family.

5. Chemokine receptor family.

There are three subfamilies of Class I cytokine receptors, which are based on the presence of an identical signal-transducing subunit. The first subfamily is GM-CSF receptor subfamily, which includes receptors for IL-3, IL-5 and GM-CSF. These cytokines bind to low affinity receptors which have a common β subunit that is the signal transducing subunit, and an α subunit that associates noncovalently to β. Due to the common β subunit, IL-3, IL-5 and GM-CSF show redundancy. The second subfamily is the IL-6 receptor subfamily, which has got a similar setup. This subfamily has receptors for IL-6, IL-11, Leukaemia Inhibiting Factor (LIF), OncoStatin M (OSM) and Ciliary Neurotrophic Factor (CNTF). The common signal-transducing subunit is gp130, which associates with 1 or 2 different cytokine-specific subunits. The third subfamily is the IL-2 receptor subfamily, which has a third signal-transducing subunit. This subunit includes receptors for IL-2, IL-4, IL-7, IL-9, IL-15 and IL-21. IL-2 and IL-15 receptors are heterotrimers that have a cytokine-specific α chain, and two chains, β and γ, for signal transduction.

15.2.2 Cytokine Secretions of T_h-cell Subsets

The cytokines that are produced by T_h subsets have very distinctive functions. T_h1 derived cytokines elicit cell-mediated immune functions, whereas T_h2 derived cytokines promote humoral response.

Cytokine secretion patterns determine the type of immune response to a particular antigen. T_h1- and T_h2-cells can be distinguished by the cytokines that they secrete. However, both the types secrete IL-3 and GM-CSF.

T_h1 secretes IFN-γ that activates macrophages. There is thus, an increased microbial activity. IFN-γ also upregulates the level of Class II MHC molecules that induces class switching to IgG that support phagocytosis and complement fixation. The TNF-β and IFN-γ both mediate inflammation, while IL-2 and IFN-γ, secreted by these cells, promote differentiation of fully cytotoxic Tc-cells from $CD8^+$ precursors.

T_h2 secrete IL-4 and IL-5 that help in the production of IgE, and supports attack mediated by eosinophils on helminth infections. IL-4 also helps in class switching to IgG that does not activate complement. The switch from IgM to IgE is mediated by IL-4. IL-4 is also needed for the development of T_h2 response.

Cross regulation

The cytokines produced by T_h1 and T_h2 promote the growth of the subset that produces them and inhibit the development and activity of the other subset. Such cross regulation accounts for the inverse relationship between antibody production and cell-mediated immunity.

15.3 PROPERTIES OF CYTOKINES

The secretion of cytokines is for short durations. Half-life of the secreted cytokines is very short, ensuring that they act for a limited period and thus, over a short distance. They are not preformed and stored, but are formed by new gene transcription.

Cytokines show different types of properties that include:

Pleiotropy: A given cytokine has different biological effects on different target cells.

Redundacy: Two or more cytokines may mediate similar functions. This makes it difficult to ascribe a particular activity to a single cytokine.

Synergism: When the combined effect of two cytokines (on cellular activity) is greater than the additive effects of the individual cytokines.

Antagonism: Effect of one cytokine inhibits the effects of another cytokine.

Cascade induction: Action of one cytokine on a target cell induces that cell to produce one or more other cytokines, which may in turn induce other target cells to produce other cytokines.

15.4 FUNCTIONS OF CYTOKINES

Cytokines mediate biological effects at picomolar concentrations. They show *autocrine action* when cytokines bind to receptors on the membrane of the cell that secreted it. *Paracrine action* is seen when cytokines bind to receptors of target cells in close proximity of the producer cell. *Endocrine action* is effected when cytokines bind to receptors of target cells in distant body parts.

The requirement of cytokine involvement is seen in the following ways:

1. They act as mediators of natural immunity.
2. They act as stimulators and regulators of lymphocyte activation, growth and differentiation.
3. Cytokines help in the mediation and regulation of immune and inflammatory response.
4. Some cytokines act as growth factors and stimulate mitosis of target cell.
5. Cytokines induce signal transduction and gene expression.
6. They help in development of humoral and cellular responses
7. Act as regulators of haematopoiesis
8. Some cytokines help in healing of wounds.

15.4.1 Cytokines that Stimulate Haematopoiesis

The cytokines that are produced during natural immunity and antigen-induced specific immune response have strong stimulatory effect on bone marrow stem cells. The progressive division and differentiation of pluripotent stem cells is completely dependent on cytokines, which bring about maturation of haematopoietic cells and commit them to a particular lineage. The responsible cytokine is called **Colony Stimulating Factor (CSF)**. Different types of CSFs act at different stages of development of bone marrow cells and promote development of cells of different lineage. Some of the actions of CSFs are influenced by other cytokines such as, Tumour Necrosis Factor (TNF), Leukotrienes (LT), Interferon-γ(IFN-γ), and Tissue Growth Factor-β (TGF-β). All of these inhibit growth of bone marrow progenitor cells.

Interleukin-3 (IL-3)

It is also called multilineage Colony Stimulating Factor (multi-CSF) and acts on the most immature bone marrow progenitor cells and promotes the expansion of all cell types. The receptor consists of a WSXWS unit and a 150-kDa signal-transducing unit shared with IL-5 and GM-CSF. IL-3 promotes the growth of mast cells derived from bone marrow progenitors.

Granulocyte-Macrophage Colony Stimulating Factor (GM-CSF)

GM-CSF is a 22-kDa glycoprotein secreted by activated T-cells and mononuclear phagocytes, vascular endothelial cells and fibroblasts. Its receptor and the signal-transducing unit are similar to the ones for IL-3. In humans, GM-CSF acts on non-committed form of leukocytes as well as platelet cells and red blood cells. It also acts as an activator of macrophages. Since it is not detected in circulation its action is mostly local, that is, bone marrow or at the site of inflammation by mononuclear phagocytes, vascular endothelial cells and fibroblasts.

Monocyte-macrophage Colony Stimulating Factor (M-CSF)

Also called CSF-1, it is secreted by endothelial cells, macrophages and fibroblasts. The secreted polypeptide is a dimer of approximately 40-kDa. M-CSF acts on progenitors that are already committed to develop into monocytes.

Granulocyte Colony Stimulating Factor (G-CSF)

Activated T-cells and mononuclear phagocytes, vascular endothelial cells and fibroblasts secrete G-CSF. The secreted polypeptide is 19-kDa and its receptor is made up of WSXWS unit. It acts on progenitors of bone marrow, which are mature and already committed to develop into granulocytes. It is found in circulation and brings about neutrophil maturation and release from bone marrow during inflammatory reaction occurring outside the bone marrow.

Interleukin-7 (IL-7)

IL-7 is secreted by bone marrow stromal cells and acts on haematopoietic progenitors committed to the B-lymphocyte lineage. It also stimulates the growth and maturation of immature $CD4^-CD8^-$ T-cell precursors in the thymus. IL-7 is required for T- and B-precursor cell maturation and survival, $CD8^+$ thymocyte positive selection, and regulation of T-cell receptor γ. Although IL-7 is a potent stimulator of immature B-cell expansion, it alone is not sufficient to promote precursor T-cell division.

15.4.2 Cytokines that Mediate Innate Immunity

Those cytokines that protect against viral infection, and initiate inflammatory reactions for protection from bacterial infection come under this class.

Type I Interferons (IFN)

Two distinct groups of proteins known as α and β fall under this type. Mononuclear phagocytes or macrophages, fibroblasts, NK-cells, T-cells, dendritic cells and the group of specialized leukocytes called **plasmacytoid monocytes**, produce these interferons. There is very less structural similarity between IFN-α and IFN-β, but both bind to same cell surface receptor and induce similar cellular responses. The biological functions of these interferons include:

(i) Inhibition of viral replication
(ii) Inhibition of cell proliferation
(iii) Increasing the lytic potential of NK-cells
(iv) Modulating MHC expression
(v) Suppression of IL-12 production
(vi) Prevention of apoptosis and promotion of IL-15 synthesis

Tumour Necrosis Factor (TNF)

TNF is secreted by the lipopolysaccharide-activated-mononuclear phagocytes and activated T-cells, NK-cells and activated mast cells. It is considered a mediator of natural and acquired immune response. The biological function of TNF is concentration dependent. At low concentration (10^{-9} M), TNF acts locally as paracrine and autocrine regulator of leukocytes and endothelial cells. At low concentrations TNF shows the following functions:

- Expression of adhesion molecules on the vascular endothelial cell surface so that it becomes adhesive for leukocytes. The TNF also increases the aggressiveness of neutrophils for endothelial cells
- Activation of neutrophils, eosinophils and mononuclear phagocytes to increase their ability to kill the microbes that cause inflammation
- Stimulation of mononuclear phagocytes to produce cytokines like IL-1, IL-6, TNF and chemokines
- Augmenting the expression of Class I MHC molecules and exerting protective effect on cells

The systemic actions of TNF are as follows:

- The TNF acts as a pyrogen that acts on hypothalamic thermoregulatory centres and induces fever
- Induction of secretion of IL-1 and IL-6 by mononuclear phagocytes
- Enhancement of production of serum amyloid A protein by hepatocytes
- Alteration of the delicate balance between anticoagulant and procoagulant activities of vascular endothelium
- Being a strong inhibitor of bone marrow stem cell division, TNF may lead to lymphopenia or immunodeficiency

Lethal effects of TNF at extremely high concentrations are as follows:

- Intravascular thrombosis due to activation of neutrophils
- Relaxation of vascular smooth muscle tone and therefore, reducing blood pressure
- Depression of myocardial contractility by involving enzymes like nitric oxide synthase
- Causing hypoglycemia by over utilization of blood glucose by muscle cells

Interleukin-1 (IL-1)

Mononuclear phagocytes secrete IL-1 which is the principal mediator of host inflammatory responses in natural immunity. At low concentrations IL-1:

- Induces the synthesis of IL-6 and IL-1 by vascular endothelial cells and mononuclear phagocytes
- Acts as a mediator of local inflammation
- Induces secretion of adhesion molecules
- Causes mononuclear phagocytes and endothelial cells to synthesize chemokines that act as activators of leukocytes

At high concentrations the biological effects of IL-1 are as follows:

- Exerting endocrine effect and causing fever
- Induction of production of serum amyloid A protein by hepatocytes and initiating metabolic wasting

Interleukin-6 (IL-6)

It is synthesized by mononuclear phagocytes, vascular endothelial cells, fibroblasts and other cells in response to IL-1. IL-6 induces hepatocytes to produce certain acute-phase proteins like fibrinogen. It serves as the growth factor for activated B-cells and of early haematopoietic stem cells of the bone marrow.

Chemokines

Chemokines have important roles to play in infectious diseases, in haematopoiesis and lymphocyte development, in cell migration, in the secondary lymphoid organs, in dendritic cell migration and T-cell activation. They also have an effect on differentiation and homing of effector lymphocytes, as well as in autoimmunity.

15.4.3 Cytokines that Regulate Lymphocyte Activation, Growth and Differentiation

Interleukin-2 (IL-2)

It is also known as T-Cell Growth Factor (TCGF) and is a major cytokine that stimulates lymphocyte progression from G1 to S phase of cell cycle. It is produced by $CD4^+$ and $CD8^+$ T-cells and acts as an autocrine factor for $CD4^+$ cells. Other cells that produce IL-2 are T-cells, FcεRI+-cells, mast-cells, basophils and eosinophils. IL-2 stimulates the growth of NK-cells and enhances their killing function, thus changing them into Lymphokine Activated Killer (LAK) cells. IL-2 acts in synergy with IL-12 to induce INF-γ secretion by NK-cells. It also acts as a B-cell growth factor and stimulates antibody synthesis (Figure 15.1).

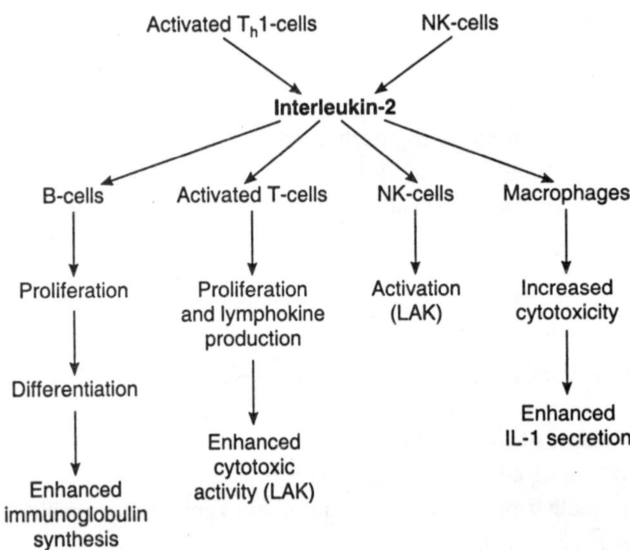

Figure 15.1 A schematic representation showing the origin and targets of IL-2.

Interleukin-4 (IL-4)

IL-4 has an important role to play in antibody production, haematopoiesis, inflammation and development of effector T-cell responses. The principal functions of IL-4 are B-cell activation, influencing cell differentiation and affecting cells that contribute to inflammatory responses. It has a central role in the development of allergic responses, in tumour immunity and in autoimmunity (Figure 15.2).

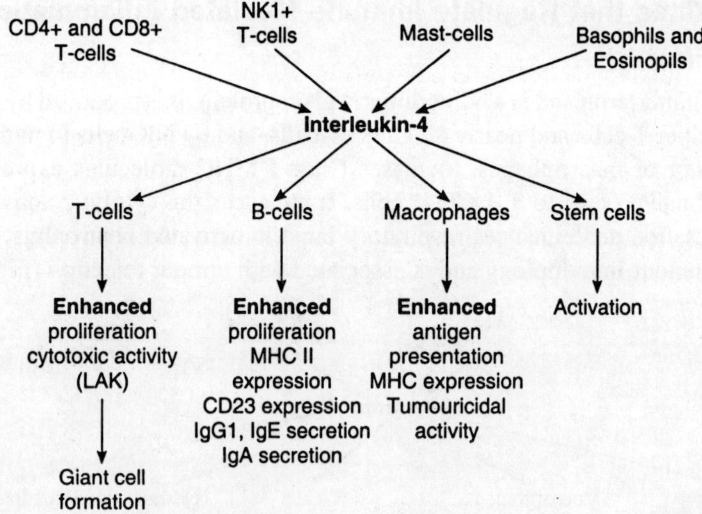

Figure 15.2 A schematic repersentation showing the origin and targets of IL-4.

Transforming Growth Factor-β (TGF-β)

It is a glycoprotein that plays a very crucial role in embryonic development, tumour genesis, wound healing, fibrosis and immunoregulation. It is produced by platelets, activated macrophages, B-cells and T-cells in an inactive form, which is later activated by proteases before it binds to the receptor. TGF-β is stimulatory to fibroblast, monocyte and neutrophil chemotaxis. It has inhibitor effects on IL-1R expression, respiratory burst, and cytotoxic effects of activated macrophages. It also inhibits monocyte and T-cell, and B-cell differentiation (Figure 15.3).

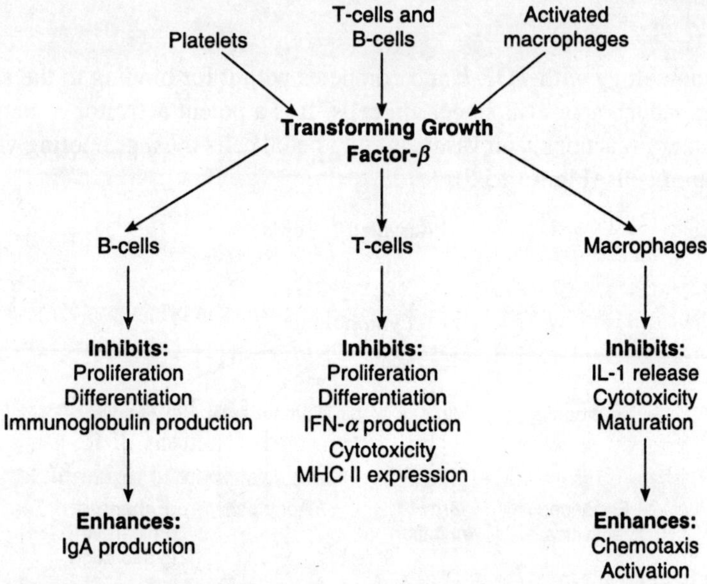

Figure 15.3 A schematic representation showing the origin and targets of transforming growth factor-β.

15.4.4 Cytokines that Regulate Immune Mediated Inflammation

Interferon-γ (IFN-γ)

IFN-γ is a Type II interferon and is a homodimeric glycoprotein. It is produced by both naive (T_h0) and T_h1 $CD4^+$ helper T-cells and nearly all $CD8^+$ T-cells, and by NK-cells in nude mice. IFN-γ is the potent activator of macrophages, increases Class I MHC molecules expression, promotes differentiation of naive (T_h0) to T_h1 $CD4^+$ cells. It enhances the cytolytic activity of NK-cells, promotes extravasation and enhances respiratory burst in activated neutrophils. IFN-γ is also a key cytokine in tumour immunology and is associated with tumour rejection (Figure 15.4).

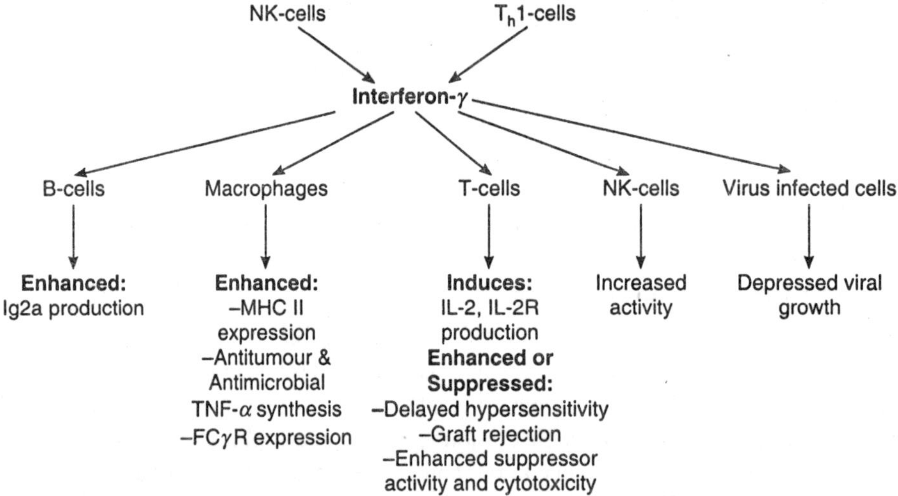

Figure 15.4 A schematic representation showing the origin and targets of gamma interferon.

Lymphotoxin (LT)

It has structural homology with TGF-β and competes with it for binding to the same cell surface receptors. LT can induce apoptosis in certain cells. It is a potent activator of neutrophils helping them in inflammatory reactions, and vascular endothelial cells thus, promoting vascular adhesion and extravasation of cells (Figure 15.5).

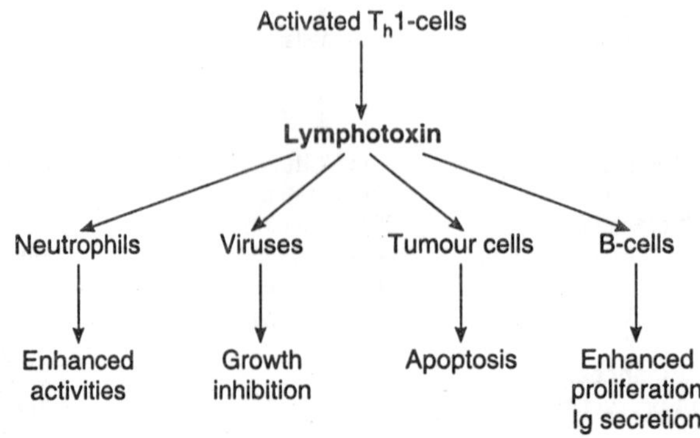

Figure 15.5 A schematic representation showing the origin and targets of lymphotoxin.

Interleukin-5 (IL-5)

IL-5 is produced by T_h2 subset of $CD4^+$ cells and by activated mast cells. It enhances the production of IgA by B-cells. Acting synergistically with IL-2, it induces cytotoxic T-cell activity (Figure 15.6).

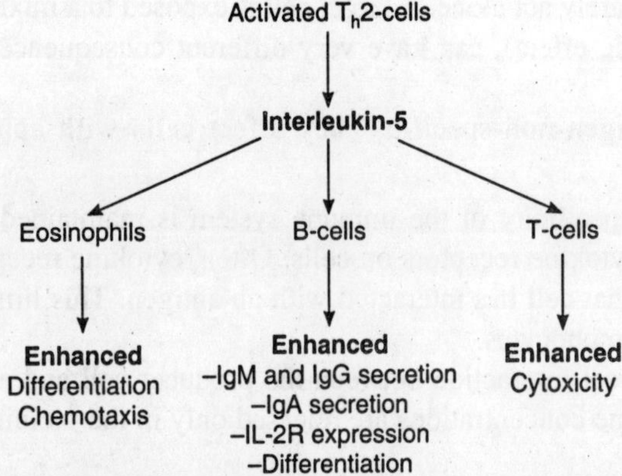

Figure 15.6 A schematic representation showing the origin and targets of IL-5.

Interleukin-10 (IL-10)

In humans, IL-10 is produced both by T_h1- and T_h2-cells and down regulates both cell types. It is also secreted by some B-cells and by activated macrophages. IL-10 acts on activated macrophages to suppress a secretion of IL-1, IL-12, TNF-α and free radicals of oxygen (Figure 15.7).

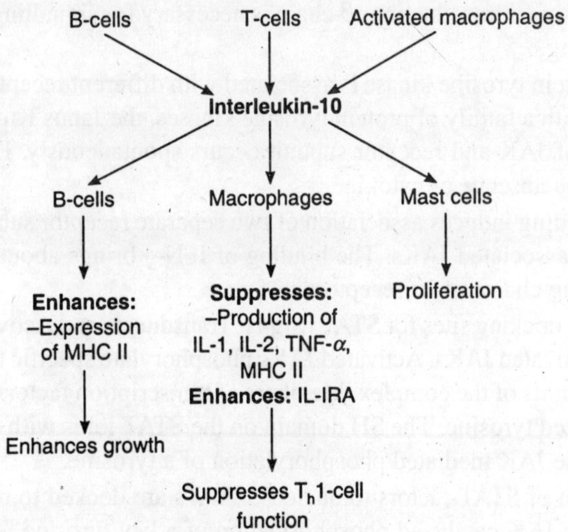

Figure 15.7 A schematic representation showing the origin and targets of IL-10.

Interleukin-12 (IL-12)

It is the 70-kDa heterodimer polypeptide made up of two covalently linked chains of 40-kDa and 35-kDa. T- and B-cells, NK-cells and monocytes produce the 35 kD chain, whereas 40 kD chain is produced only by activated monocytes and B-cells. IL-12 is an important regulator of cell-mediated immune responses. It stimulates NK-cells and induces transcription of IFN-γ by NK-cells. It enhances cytolytic activity of NK-cells, as well as acts like a growth factor for these cells. It stimulates the differentiation of $CD8^+$ T-cells into mature and functionally active CTLs.

In vivo, cytokines rarely act alone. A target cell is exposed to a mixture of cytokines (whose synergistic or antagonistic effect), can have very different consequences. Cascading activity is also seen.

Cytokines are antigen-non-specific. They affect cells with appropriate receptors and physiological state.

1. One way that specificity of the immune system is maintained is by regulation of the expression of cytokine receptors on cells. Often, cytokine receptors are expressed on a cell only after that cell has interacted with an antigen. This limits cytokine response to Ag-activated lymphocytes.
2. Moreover, a direct interaction between the producer cell and a target cell ensures that effective cytokine concentrations are released only in the vicinity of the target cell.

15.4.5 Signal Pathway Activated by Cytokine Receptors

Studies have been done on binding of γ-IFN to its receptor belonging to Class II family, and a unifying cytokine-signalling model has emerged. γ-IFN regulates the mononuclear phagocytes, B-cell switching to IgG classes, and inhibition or development of T_h-cell subsets. Signal pathway is activated by cytokine receptors in the following manner:

1. Presence of separate subunits in the cytokine receptor; α-chain is required for cytokine binding and signal transduction; β-chain is necessary for signalling, but has only a minor role in binding.
2. Inactive protein tyrosine kinase is associated with different receptor subunits. α-chain is associated with a family of protein tyrosine kinases, the Janus Kinase (JAK) family. The association of JAK and receptor subunit occurs spontaneously. However, JAK remains inactive in the absence of cytokine.
3. Cytokine binding induces association of two separate receptor subunits and activation of the receptor associated JAKs. The binding of IFN-γ brings about the association of the ligand binding chains of its receptors.
4. Formation of docking sites for STAT (Signal Transducers and Activators of Transcription) factors by activated JAKs. Activated JAKs phosphorylate specific tyrosine residues in the receptor subunits of the complex. Members of transcription factors (STATs) bind to these phosphorylated tyrosine. The SH domain on the STAT joins with the docking site that is created by the JAK mediated phosphorylation of a tyrosine.
5. Translocation of STAT factors to nucleus. STATs are docked to receptor subunits. Here they undergo JAK-catalyzed phosphorylation of a key tyrosine. Then STATs dissociate from the receptor and dimerize. The dimers translocate into the nucleus to induce the expression of genes containing regulatory sequences (Figure 15.8).

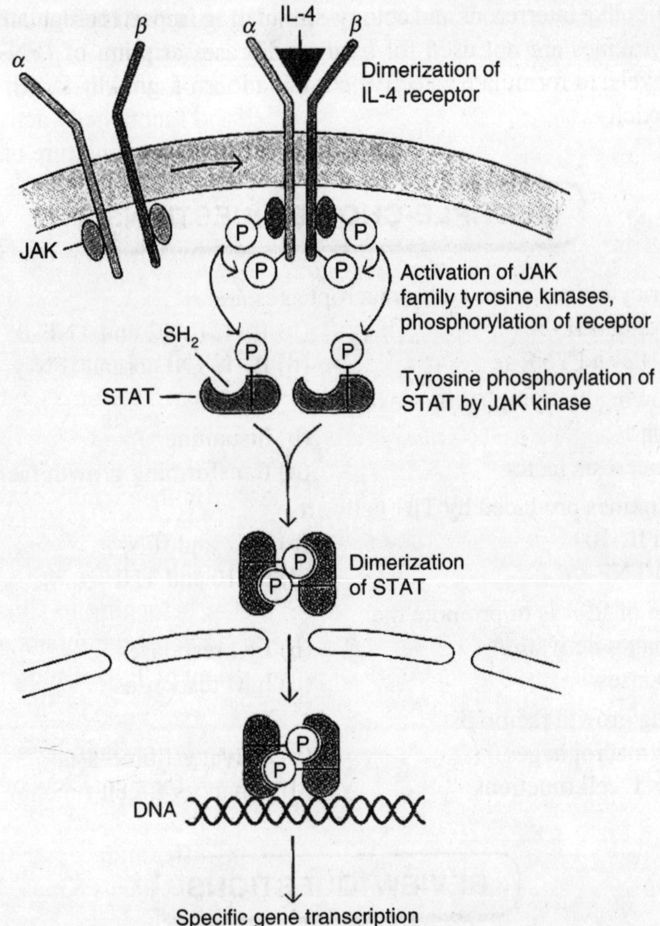

Figure 15.8 Signal pathway activated by cytokine receptors.

15.5 CYTOKINE ANTAGONISTS

A number of proteins inhibit the biological activity of cytokines by either

1. Binding directly to a cytokine and inhibiting its activity, or
2. Binding directly to a cytokine receptor, but fail to activate the cell.

Some viruses have developed strategies to stop cytokine activity by producing cytokine homologs, soluble cytokine-binding proteins and homologs of cytokine receptors. They also interfere with cellular signalling and cytokine secretion. Moreover, they also induce cytokine inhibitors in the host cell.

Any abnormality in the expression of cytokines or their receptors may result in diseases like bacterial toxic shock, some lymphoid and mycloid cancers, and Chagas' disease. Cytokine-related therapies offer promise of reducing graft rejection, treating some cancers and immunodeficiency diseases, and reducing allergic reactions, by increasing or decreasing immune response. Therapeutic use of some cytokines like interferons and colony-stimulating factors is a standard medical practice. However, many cytokines are not used for treating diseases as there is a need to first maintain appropriate dose levels, to formulate effective combinations of cytokines with other drugs and to control cytokine toxicity.

MULTIPLE-CHOICE QUESTIONS

1. Three major cytokines secreted by macrophages are:
   (a) IL-1, IL-2 and IL-3 (b) IL-1, IL-12 and TNF-α
   (c) IL-2, IL-12 and TNF-α (d) IL-1, TNF-α and IFN-γ
2. All the following are lymphokines except:
   (a) interferon (b) histamine
   (c) tumour necrosis factor (d) transforming growth factor
3. The key cytokines produced by Th1 cells are:
   (a) IL-4 and IL-10 (b) IL-2 and IFN-γ
   (c) IL-1 and TNF-α (d) IL-10 and TNF-α
4. A major role of IL-4 is to promote the:
   (a) macrophages activation (b) IgG responses
   (c) IgE responses (d) IgM responses
5. Transforming growth factor-β:
   (a) activates macrophages (b) activates fibroblasts
   (c) enhances T-cell functions (d) enhances B-cell functions.

REVIEW QUESTIONS

15.1 Write notes on:
(i) Cytokines that regulate haematopoiesis
(ii) General properties of cytokines
(iii) Cytokine receptors
(iv) Tumour necrosis factor
(v) Cytokine antagonists

15.2 What are chemokines? Explain the properties and functions of chemokines.

15.3 How is signal pathway activated by cytokines?

16

IMMUNITY TO INFECTIOUS AGENTS

Every year, infectious diseases are the cause of millions of deaths in developing countries. It is the function of the immune system to defend the body against invasion by infectious agents. There are two major groups of invaders that the body must defend itself against. The first group is the foreign substances or cells that invade the body from outside. This group includes bacteria, viruses, fungi, protozoa and helminthes. The second group includes the abnormal cells that arise inside the body, such as chemically modified cells, virus infected cells and tumor cells. People who are deficient in antibodies suffer from bacterial diseases, whereas those who are deficient in T cells suffer from recurrent viral diseases.

Resistance to diseases can result from species insusceptibility. For example, *Neisseria gonorrhoeae* grows well only in humans. Diphtheria is caused by *C. diphtheria* in humans and guinea pigs but not in rats. Non-immunological factors that influence disease resistance are genetics, age and nutrition. The very young and the very old are more susceptible to infection. For instance, rubella infections can cause severe damage to the foetus but scarcely affects the mother. Similarly, *Toxoplasma gondii* cause severe damage to the foetus, but produces only a mild infection in adults. Age related resistance can also be under the influence of hormones. Infections in the vagina are more commonly seen before puberty and after menopause. Severe malnutrition too impairs immunoglobulin production. There can be increases in protein catabolism after surgery, which can result in temporary immunosuppression. This too increases susceptibility to infections.

Many pathogens have the ability to change their antigenic markers; some are able to grow within the host cell, by reducing their antigenicity and thus remain protected from immune attack. Some pathogens shed their antigens to avoid immune detection, while others can mimic host cell membrane antigens either by making them or acquiring them from the host cell membrane. Many organisms induce hypersensitivity reactions that can result in tissue damage and increased severity of disease.

16.1 IMMUNITY TO BACTERIA

Bacteria have access to a body through respiratory tract, GI tract, genitourinary tract, damaged or broken mucous membrane or the skin. Usually, a body raises a humoral immune response to bacterial infections; however the intracellular infection may cause delayed type of hypersensitivity. A non-specific defense usually takes care of low virulence bacteria, while a strong specific immune response is evoked against a virulent strain and high scale infection.

Mechanisms of immunity are related to bacterial mechanisms of pathogenicity. There are two extreme patterns of pathogenicity; these are toxicity without invasiveness and invasiveness without toxicity. *Corynebacterium diphtheria* and *Vibrio cholerae* are examples of organisms that are toxic but not invasive. Their pathogenicity depends entirely on toxin production, and so, neutralizing antibody to the toxin is probably sufficient for immunity. Antibody binding to the bacteria and hence hindering its adhesion to the epithelium is also important. On the other hand, the pathogenicity of most organisms does not rely so heavily on a single toxin and so immunity requires killing the organism itself.

Immunity to bacteria may involve antibodies, complement or activated macrophages. Bacteria may be opsonized, lysed by complement or destroyed by activated macrophages (Figure 16.1).

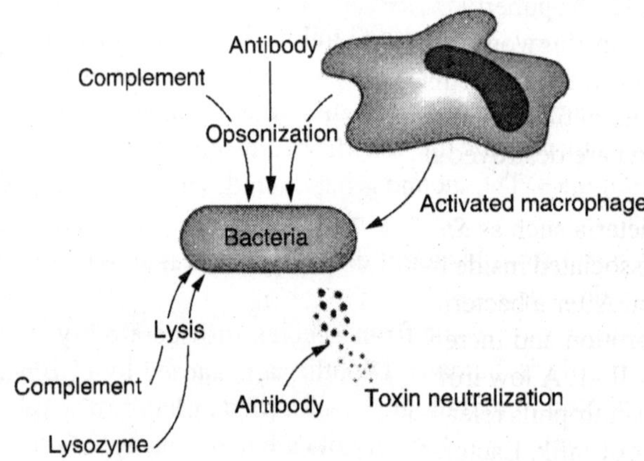

Figure 16.1 The different ways in which the immune system can protect the body against bacterial invasion.

16.1.1 Non-specific Immunity

Intact skin is impenetrable to most bacteria. Minor skin wounds do not always result in bacterial infections. This can be taken as evidence that animal tissues are able to discourage bacterial invasion. This resistance may partly be due to the presence of antibacterial factors in tissues (Table 16.1). One such factor is lysozyme which is present in tissues and all body fluids with the exception of cerebrospinal fluid, sweat and urine. Lysozyme is present in high concentrations in the lysosomes

Table 16.1 Factors of Innate Immunity found in the Body Fluids and Tissues

Group	*Name*	*Major sources*	*Activity against*
Enzymes	Lysozyme	Serum, leukocytes	Gram positive and Gram negative bacteria; some viruses
Iron-binding proteins	Transferrin Lactoferrin	Serum Leukocytes, milk	Gram positive and Gram negative bacteria
Basic amines	Spermine, spemidine	Pancreas, kidney, prostate	Gram positive bacteria
Complement components	C3a, C5a, etc.	Serum	Bacteria, viruses, protozoa
Peroxide generating mechanisms	Xanthine oxidase myeloperoxidase	Milk, neutrophils	Bacteria, viruses, protozoa
Basic peptides and proteins	B-lysin Phagocytin Leukin Plakin	Platelets Neutrophils Neutrophils Platelets	Gram positive bacteria
Interferons	TNF, IFN-α,-β,-γ,-ω	Most cells, with the exception of neutrophils	Viruses, some intracellular protozoa

of the neutrophils, and so it is found in substantial amounts at areas of inflammation and sites of bacterial invasion. Lysozyme splits up the proteoglycans of some gram +ve bacteria thus killing them. It can also kill some gram –ve bacteria with the help of complement. Moreover, fatty acids produced by skin are toxic to many organisms. Free fatty acids can inhibit bacterial growth. Unsaturated fatty acids such as oleic acid are bactericidal for gram +ve organisms, while saturated fatty acids are fungicidal. At puberty the amount of fatty acids in the sebum increases which helps in the resolution of scalp ringworm in children. An antibacterial polypeptide β-lysin is released from platelets when they interact with immune complexes. This polypeptide is active against *Bacillus anthracis.* Several other antibacterial peptides and proteins rich in lysine and arginine have been isolated. Many bacteria are destroyed by pH changes in the stomach and vagina, where an acidic pH prevails.

A number of bacteria such as *Staphylococcus aureus*, *E. coli*, *M. tuberculosis* need iron for their growth. Iron is associated inside the body with iron binding proteins transferrin, haptoglobin, lactoferrin and ferritin. After a bacterial invasion serum iron levels drop as a result of a reduction in intestinal iron absorption and increased production of haptoglobin and transferrin by the liver under the influence of IL-1. A low iron level hinders the bacterial invasion. Similarly, in response to bacterial invasion, neutrophils release their stores of lactoferrrin in mammary glands, enhancing the bactericidal power of milk. Lactoferrin is also implicated in the killing mechanism as it binds iron and makes it unavailable to the bacteria. In spite of binding the iron, some bacteria such as *M. tuberculosis* and *E. coli* are able to invade the body because they produce potent iron-binding molecules known as *siderophores*. Siderophores have a very high affinity for iron and can withdraw iron from the serum proteins to make them available to the bacteria.

16.1.2 Specific Immunity

Bacterial invasion can be fought specifically through four basic mechanisms. These are the neutralization of toxins or enzymes by antibodies; killing the bacteria by complement, antibodies and lysozyme; opsonization and phagocytosis of bacteria by neutrophils and normal macrophages;

and phagocytosis and killing the intracellular bacteria by activated macrophages. Defense is also mediated through recognition of common bacterial components. Lymphocyte—independent bacterial recognition pathways have several results. There can be activation of the complement by the alternative pathway, which may kill some bacteria and also release the chemotactic products C3a and C5a. These cause smooth muscle contraction and mast cell degranulation as well as attract and activate neutrophils. Phagocytes are also attracted to the site of infection. There is also a rapid release of cytokines from macrophages and natural killer cells.

Protection from the primary bacterial infections is quite complex; however secondary infections are usually taken care of by antibodies.

Antibody provides antigen specific protective mechanism. Antibody plays a crucial role in dealing with bacterial toxins. It neutralizes diphtheria toxin by blocking the attachment of the binding portion of the molecule to its target cells. It also blocks locally acting toxins or extracellular matrix-degrading enzymes. Secretory IgA stop bacterial binding to epithelial cells.

Bacterial exotoxins can be readily neutralized by antibodies through the blockage of the complexed toxin to its receptor on the host cell. Specific antibodies (antitoxins) neutralize the exotoxins and phagocytes remove endotoxins of the bacteria and toxin-antibody complexes. IgG is almost totally responsible for toxin neutralization.

Protection against invasive bacteria is mediated by antibodies that are directed against their surface antigens. These antibodies are able to neutralize the antiphagocytic properties of the capsule, and opsonize the bacteria thus aiding in their destruction by phagocytosis. IgG is a more effective opsonin than IgM in the absence of complement. IgM is more effectively able to activate the complement and can opsonize better in the presence of complement. A localized inflammatory reaction may also develop as a result of antibody mediated activation of complement.

Most of the bacteria are killed by phagocytes. The killing mechanism may be oxygen dependent which involves the activity of reactive oxygen intermediates or reactive nitrogen intermediates; the killing mechanism may also be oxygen-independent, which makes use of cationic proteins or lysosomal enzymes.

Some bacteria are phagocytosed by neutrophils or macrophages; others are killed while they are free in circulation. The bactericidal activity of the serum is mediated by antibodies, complement and lysozyme. Lysozyme can act on the exposed inner cell membrane following complement activity killing the bacteria. Gram negative organisms may activate the alternate complement pathway leading to their lysis. Nearly 25 new proteins are produced by stressed cells. The stress may be caused by heat, exposure to oxygen radicals, starvation, viral infections, and heavy metals. Heat shock proteins (HSPs) are one of these newly produced proteins. HSPs are present in low concentrations at normal temperatures, but their levels increase at higher temperatures. These proteins control the folding of other proteins, so that the organism can function normally at higher temperatures. There are three major HSPs: HSP60, HSP70 and HSP90. During a respiratory burst that follows infection by bacterium large quantities of HSP60 are made by the stressed organism. HSP60 is the immunodominant antigen induced by *Mycobacterium, Borrelia, Legionella* and *Treponema* infections. The gene for HSP70 is located within the MHC class III region in mammals. HSP70 is seen in patients of malaria, leishmaniasis, filariasis and trypanosomiasis. All the three HSPs are highly antigenic because they are produced in abundance, are readily processed by antigen processing cells and because there may be large numbers of lymphocytes that can respond to HSPs.

The course of infection is dictated by the development of an antibacterial immune response. At best, the infection is cured. However if it is not cured, the disease may be modified depending upon whether a humoral or cell-mediated response had occurred. Cell mediated responses can control intracellular bacteria through activated macrophages. Activation of macrophages is dependent upon the production of IL-2 and IFN-ϒ by the Th1 cells. Activated macrophages can localize and cure these infections. However, if Th2 cells are activated, it does not result in the activation of macrophages, and chronic progressive disease results as in the case of leprosy. Generally, delayed type of hypersensitive reaction occurs in infections by intracellular bacteria, where cytokines secreted by T_{DTH} cells play an important role, which includes IFN-ϒ that activates macrophages.

The response to bacteria can result in immunological tissue damage. Excessive cytokine release can lead to endotoxin shock. Endotoxin (septicemic) shock occurs when there is a massive production of cytokines, usually caused by bacterial products released during septisemic episodes. Endotoxin lipopolysachharide from Gram positive and Gram negative bacteria cause septicemia. There can be life-threatening fever, circulatory collapse, diffuse intravascular coagulation and hemorrhagic necrosis; eventually leading to multiple organ failure.

Bacteria have evolved mechanisms to evade the immune responses of the host. Some bacteria like *E. coli* and *M. tuberculosis* secrete molecules that can depress phagocytosis by neutrophils. Others like *Bordetella pertussis* secrete an adenylate cyclase that raises the levels of intracellular c-AMP and thus suppress many cellular activities. The infected people thus become susceptible to other infections. *Shigella flexinera* induces apoptosis in infected macrophages. *Pasturella haemolytica* secretes a protein that kills alveolar macrophages. *P. aeruginosa* secretes a protein that destroys IL-9. Many bacteria defend themselves against oxygen radicals produced by phagocytic cells. They can use enzymes such as superoxide dismutase and catalase, DNA repair system and give competition to phagocytes for molecular oxygen as seen in *S. aureus*. *E. coli* carries plasmids that confer resistance to complement-mediated lysis. The secretion of protective factors like proteolytic enzymes is used by *N. meningitis* and *S. pneumonia* to evade the immune responses. Endocytosed bacteria liable to be killed by lysosomal enzymes take steps to avoid being killed by the cell (Figure 16.2). They do so by inhibiting both acidification and fusion between bacteria-containing phagosomes and lysosomes. *M. tuberculosis* can survive in the phagolysosome because they are resistant to lysis by lysosomal enzymes. Another survival technique is practiced by intracellular bacteria like *L.monocytogenes*. It escapes from the phagolysosome into the cytoplasm

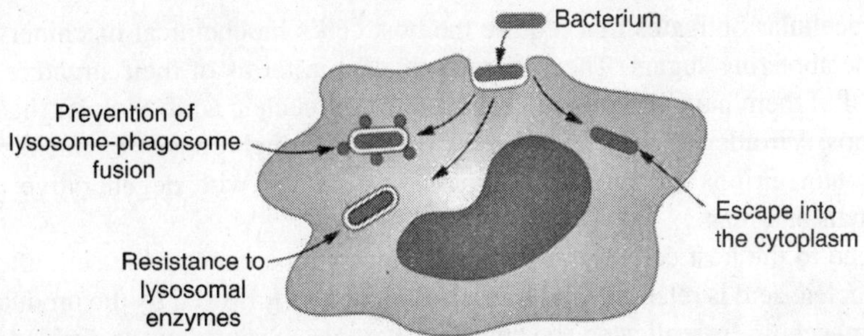

Figure 16.2 The three major ways by which bacteria and other intracellular parasites can escape destruction within macrophages.

of the host's cell. Some bacteria have the capability to interfere with the complement system. *Pseudomonas* secretes the enzyme elastase which can inactivate C3a and C5a complement components. This fails to produce an inflammatory reaction at the site of infection. Some Gram +ve bacteria produce long side chains on lipid A moiety of the cell wall-core polysaccharide, which helps to resist complement-mediated lysis.

16.2 IMMUNITY TO FUNGI

Very little is known about the precise mechanism involved in immunity to fungal infections. They are thought to be similar to those involved in resistance to bacterial infections. The fungal infections that invade humans can be categorized in the following four main categories:

1. Superficial mycoses, which are caused by fungi known as *dermatophytes*. They are usually restricted to the nonliving keratinized components of skin, nails and hair.
2. Subcutaneous mycoses, that are chronic nodules or ulcers in subcutaneous tissue formed after a trauma, for example, chromomycosis, sporotrichosis and mycetoma. Saprophytic fungi can cause such mycoses.
3. Respiratory mycoses, which are produced by soil saprophytes. They produce subclinical or acute lung infections, or granulomatous lesions. Examples are histoplasmosis and coccidioidomycosis.
4. Candidiasis, which is caused by *Candida albicans,* is manifested as superficial infections of skin and mucous membranes.

Cutaneous fungal infections are usually self-limiting; after recovery, there is a limited resistance to reinfection. Resistance is based on cell-mediated immunity and can be transferred with immune T cells. The T_H cells release cytokines that activate macrophages to destroy the fungi. Disturbance of natural physiology by immunosuppressive drugs, or of normal flora by antibiotics can predispose to invasion by *Candida.* Diseases caused by *Candida* are also common in persons with immunodeficiency diseases. The immune system is thus involved in confining the fungus to its normal commensal sites.

16.3 IMMUNITY TO VIRUSES

Viruses are intracellular obligates that require the host cell's biochemical machinery for protein synthesis and metabolizing sugars. They are very diverse in terms of their structure and genetic complexities. All of them have little more than protein and nucleic acid. Simpler than viruses are viroids and prions. Viroids are infectious agents of plants which consist of nucleic acid alone, encoding no protein. Prions are infectious proteins associated with degenerative neurological diseases of animals and man.

Viruses bind to the host cells through specific receptors and enter the cell, after which the virus uncoats, nucleic acid is released, and transcription occurs followed by the production of viral proteins. The viral genome is replicated and new virus particles (virions) are assembled and released to infect the neighbouring cells. The time taken for this process depends on the infecting virus and the metabolic state of the host cell. Picornaviruses (small RNA viruses) take about eight hours to produce new virions while human cytomegalovirus (a DNA virus) takes up to forty eight hours.

Viruses are very diverse in terms of infection, latent periods and persisting, and producing disease. Entry is usually at mucous surfaces. Some viruses, like the herpes virus, persist in latent (non-infectious) form after acute infection is resolved, and can reactivate to produce new infectious virions. Other viruses can persist in the infectious form despite the presence of an immune response (like hepatitis B virus).

The initial defense against viral invasion is the integrity of the body surfaces. If this is breached, the innate defenses such as interferons, natural killer cells and macrophages become active (Figure 16.3).

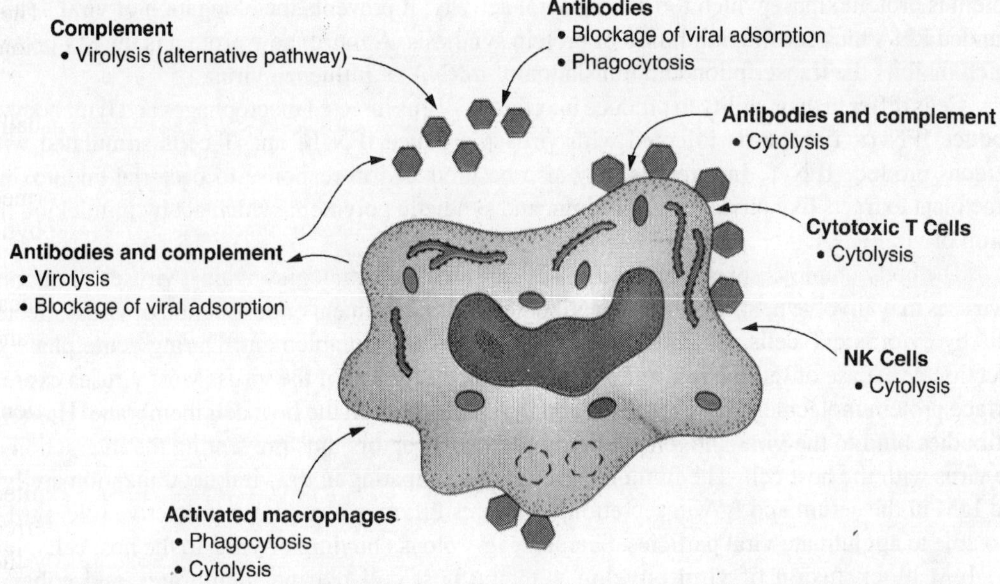

Figure 16.3 The ways in which the immune system can defend the body against viral invasion.

When viral infection occurs many specific immune effector functions (both innate and adaptive) destroy and eliminate the virus from the body. The innate system recognizes the general molecular pattern of the bacteria/virus or any post transitional modifications caused in the infected host cells. Monocytes, macrophages, eosinophils, natural killer cells and the soluble mediators like the components of complement or acute phase reactions can kill the virus directly or indirectly.

Nonimmunological factors too are able to modify and control the outcome of many viral infections, for example, lysozyme and many other intestinal enzymes can destroy several viruses. Bile is a powerful neutralizer of some viruses and protects the intestine against invasion.

Virus infected cells secrete interferons within a few hours after viral invasion, and within a few days high quantities of interferons are seen in the tissues at a time when the primary immune response is still ineffective. This is seen to happen in individuals exposed to influenza virus. The infected individual forms α and β interferon, which activate the antiviral mechanisms of neighbouring cells by activating genes with antiviral activity. An example is the Mx gene that inhibits the primary transcription of influenza virus genes, but has no effect against other viruses. Natural killer cells are also seen to be cytotoxic for virally infected cells. Interferons are made by the association between viral nucleic acid and host cell ribosomes, resulting in derepression of target cells' genes for coding interferon. Interferon is secreted by infected cells and binds to receptors of other nearby cells. This binding confers a resistance to viral infection within a few minutes and reaches a peak 5 to 8 hours later. Interferons indirectly stimulate the formation of new proteins in target cells; some of these new proteins have antiviral activity. One of these proteins is the enzyme 2′5′-oligoadenylate synthase, which plays an important role in the activation of a latent cellular endoribonuclease (RNAase L). Once activated, RNAase cleaves viral mRNA and thus inhibits viral protein synthesis. Nitric oxide synthase is another enzyme induced by INF-ϒ in the activated macrophages. The nitric acid produced by this enzyme has antiviral activity. A third induced protein is protein kinase which too has antiviral activity; it prevents the elongation of viral double stranded RNA thus causing inhibition of protein synthesis. A fourth new protein is the Mx protein which inhibits the transcription and translation of mRNA of influenza virus.

Cells differ in their ability to produce interferons. Virus infected macrophages and lymphocytes produce IFN-α; fibroblasts infected with viruses produce IFN-β; and T-cells stimulated with antigens produce IFN-ϒ. Interferons may also be produced in response to bacterial endotoxins, some plant extracts like phytohemagglutinin, and synthetic polymers, which act by mimicking the action of viral RNA.

The body employs several antibody- and cell-mediated strategies against viruses. Immunity to viruses may involve neutralization by antibodies and complement or destruction of virus-infected cells by cytotoxic T-cells, by ADCC or by antibodies and complement. During acute phase of infection or in case of reinfection, antibodies prevent the spread of the virus. Most viruses express surface protein molecules that specifically bind to molecules on the host cell membrane. Humoral antibodies bind to the viral antigenic determinants or receptors thus preventing the interaction of the virus with the host cell. The immunoglobulins participating in the viral neutralization are IgG and IgM in the serum and IgA in secretions. At times IgE may also play a protective role. IgM is also able to agglutinate viral particles. Secretory IgA blocks binding of virus to the host cells; IgG and IgM block fusion of viral envelop with the host cell plasma membrane, and enhance phagocytosis of the viral particles.

B-cells are pivotal for elimination of viruses that are structurally highly organized. All viruses can be exclusively controlled by antibodies if antibodies are present in the body before an infection occurs. Antibody is transferred from the mother to the foetus/child through blood/milk. This antibody is able to protect the new-born baby against a variety of viral infections from hepatitis B, rubellavirus, influenza virus and herpes virus. In addition, preexisting neutralizing antibodies are able to protect from viral infections that are otherwise exclusively controlled by CTLs during primary infections. Such viruses include lymphocytic choriomeningitis virus (LCMV), hepatitis B virus and primary isolates of HIV.

Antiviral antibodies are directed against the antigenic glycoproteins present on the capsid of the virus. Antibodies may destroy free viruses, prevent viral infection of cells, or destroy virus-infected cells. The antibodies may also neutralize the infectivity of viruses by blocking their attachment on the cell receptors. The T-even phages (T2, T4, etc.) have only one critical attachment site; they can thus be neutralize by blocking that site with one immunoglobulin molecule. Viruses that have multiple binding sites, such as the influenza virus, need much higher levels of immunoglobulin for their neutralization as all the sites need to be blocked. Antiviral antibodies may act as opsonins, promoting phagocytosis of virions by macrophages, by clumping the virions, or by initiating complement-mediated virus neutralization. The mobilized complement and effector cells destroy virus-infected cells. Complement is involved in the neutralization of free viruses. T-cells mediate viral immunity through the activity of $CD8^+$ cytotoxic T cells. $CD4^+$ cells also have important effector functions against virus infections.

Viruses such as vesicular stomatitis virus, polyoma virus and rotavirus exhibit highly repetitive surface antigens and therefore induce IgM antibodies in the absence of T_H cells. Protection against influenza virus needs the activity of both B and T cells. For hepatitis B virus, $CD8^+$T cells play a predominant role while $CD4^+$T cells and B cells do not have an important role. Examples of viruses exclusively controlled by T cells are lymphocytic choriomeningitis virus (LCMV) simian immunodeficiency virus (SIV) and HIV. Antibody-dependent cellular cytotoxicity is an important mechanism of destruction of virus-infected cells.

Antibodies are not able to completely eliminate the virus once the infection has occured and when the virus has entered the latent phase after incorporating its genome into the host genome. Thus once infection has occurred, it is the cell-mediated mechanism which can get rid of the infection. The T_{H1} cells are activated after coming in contact with the viral antigens, and produce cytokines such as IL-2, IFN-ϒ and TNF that act as immune mediators. IFN-ϒ induces an antiviral state in the cells, while IL-2 activates the CTLs. IL-2 and IFN-ϒ also activate the NK cells. Most of the virions are eliminated within 7–10 days of infection by CTL activity.

An important mechanism of resistance to viruses is cell mediated immunity directed against virus-infected cells. Cell-mediate responses destroy intracellular parasites; individuals with deficient cell-mediated immune responses often die of viral infections.

Virus infected cells develop new antigens on their surfaces, which include virus proteins. Cell-mediated destruction of such cells is mediated by NK cells or cytotoxic T-cells.

To evade the immune response of the host, viruses have evolved a number of mechanisms (Table 16.2).

Table 16.2 Strategies used by Viruses to Outwit the Host Defenses

Host defense effected	*Virus*	*Mechanism*
Interferon	EBV	Blocks protein kinase activation
	Vaccinia	Prevents phosphorylation of eIF-2α by protein kinase
Complement	Vaccinia	Blocks complement activation
	HSV-1	Binds Fc-ϒ and blocks function
Cytokines	Myxoma	Competes for IFN-ϒ, blocks function
	Shope fibroma virus	Competes for TNF, blocks function
	EBV	Reduces IFN-ϒ function
MHC class I	Murine cytomegalovirus	Prevents transport of peptide loaded MHC
	Adenovirus	Blocks transport of MHC to surface

Many viruses produce proteins that interfere with specific and nonspecific host responses. Others like the vaccinia virus produces vaccinia complement-control protein which promotes the cleavage of C3b and C4b by factor I, thus inhibiting the classical complement pathway. Herpes virus has a glycoprotein that can bind to C3b component of the complement, thereby inhibiting both classical and alternative pathway of complement. A number of viruses such as the influenza virus escape immune attack by often changing their antigens resulting in frequent infection with a new strain. The influenza virus in a population change their antigenic structure by changing their amino acid sequences gradually, resulting in a gradual change in the antigenicity of a subtype. This antigenic drift permits the virus to persist in a population for many years. Influenza virus also undergoes a major change in antigenic structure (antigenic shift) in which new strains develop. Such a major change is usually due to recombination between two strains of virus and accounts for the periodic major epidemics of influenza. It is seen in HIV and in foot and mouth disease virus and is responsible for antigenic shift and drift seen with influenza virus. Humoral immunity to such diseases lasts only until the new virus strain emerges. This makes effective long-lasting vaccinations difficult to produce. The degree of antigenic variation is highest in the HIV, which accumulates mutations 65 times faster than the influenza virus.

Many viruses benefit from antagonism of cytokine activity. Some viruses encode their own cytokines. Epstein Barr virus encodes a protein homologous to the cellular cytokine IL-10 which is a negative regulator of IL-12. IL-12 promotes IFN-ϒ production and has a big impact on the development of cells producing Th1 and Th2 like cytokines. Viruses may also neutralize cytokine activities; adenoviruses encode four genes whose products antagonize the effect of tumour necrosis factor (TNF).

In many cases, the antiviral immunity is long-lasting. This is related to the virus persisting within the cells in a non-replicating or slowly replicating form, as is the case of herpes virus. The persistent virus may periodically boost the immune response of the infected animal thus generating prolonged immunity. If the viral antigens are not expressed on the cell-surface, the cells are not eliminated. The immune response in such cases would not eliminate the virus but may prevent the development of disease and therefore be protective. This evasive method is seen in measles infected cells.

Under stressful conditions there is an increased production of ACTH, which may be immunosuppressive enough to permit activation of latent viruses or infection by exogenous viruses. Immunosuppression may also occur as a result of viral diseases, which may destroy lymphoid tissue. Mice infected by herpes virus show necrosis of thymus; in poultry the bursa of Fabricius is destroyed by virus of infectious bursal disease. Some viruses damage secondary lymphoid organs as in measles infection. The Epstein Barr virus infects only the B-cells and activates them, causing them to proliferate and form tumours. These activated B-cells are recognized as foreign by the cytotoxic T-cells and destroyed.

Viruses commonly suppress the expression of the MHC molecules on the infected cells, especially the MHC class I molecules. In some cases viruses are able to block the effects of interferon on MHC expression.

16.4 IMMUNITY TO PROTOZOA

Protozoan parasites are unicellular, intracellular or extracellular pathogens that cause many diseases in humans like amoebiases, African sleeping sickness and malaria.

Parasite protozoa may live in the gut (e.g., amoeba), in the blood (e.g., African trypanosomes), within erythrocytes (e.g., *Plasmodium*), in macrophages (e.g., *Leishmania*) in liver or spleen (e.g., *Leishmania*) or in muscle (e.g., *Trypanosoma*).

The type of immune response shown by the host depends on the location of the parasite within the body. Two major effector functions of natural killer cells are seen during protozoan infection: lysis of the target cells and the production of cytokines involved in the production of other immune cells. NK cells also destroy host cells like macrophages infected with *Plasmodium falciparum, T. gondii* or *Leishmania major* (Figure 16.4).

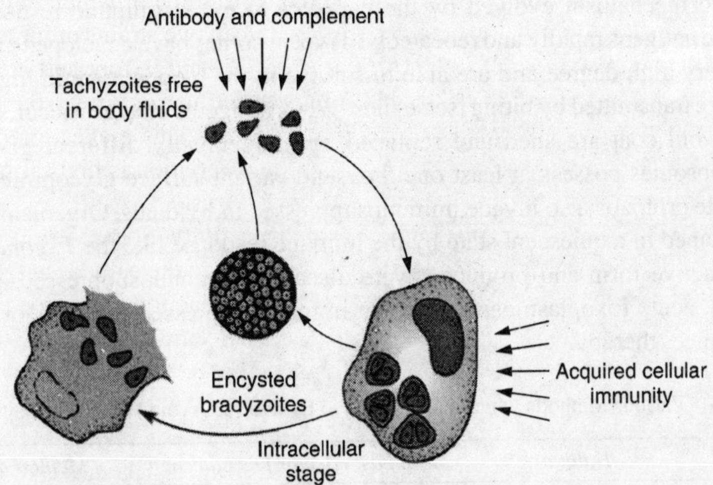

Figure 16.4 The ways in which immune responses act together to destroy protozoan parasites like *T. gondii.* (Tachyzoites are replicating forms found in the body. Bradyzoites are the forms found within cysts).

Genetically, determined resistance to protozoan disease is seen in sickle cell anemia in humans. Red blood cells containing hemoglobin S are not parasitized by *Plasmodium falciparum.* Such individuals are hence resistant to malaria. Persons who lack the Duffy blood antigens are also resistant to malaria. The membrane molecule that carries the Fy epitope is also the receptor for *Plasmodium vivax,* without which the parasite cannot invade the erythrocyte.

Most parasites are fully antigenic but they have developed mechanisms through which they are able to survive an immune response. The antibodies control the levels of parasites in the blood and tissue fluids and the cell-mediated immune responses are mediated against intracellular parasites. The production of reactive nitrogen species such as nitric oxide is important in immunity against intracellular parasites like *Leishmania.* The production of nitric oxide is promoted by TNF-α and IFN-γ. Antibodies against protozoan antigens may opsonize, agglutinate or immobilize them. Antibodies together with complement and cytotoxic cells may lyse them.

Protozoans resort to certain mechanisms to evade the host's immune response. These are as follows:

1. Immunosuppression induced by parasites may assist their survival. *Babesia bovis* is immunosuppressive for cattle. This enables its vector, the cattle tick to survive on an infected animal. *Trypanosoma cruzi* which causes Chagas' disease, prevents the expression of IL-2 receptor on T-cells and so induce immunosuppression. Parasite induced immunosuppression often leads to the death of the host because of secondary infection.
2. Another effective immunoevasive technique involves the parasite becoming 'hypoantigenic'. An example is the cyst stage of *T. gondii* that does not stimulate a

significant immune response. Some protozoans become nonantigenic by masking themselves with host antigens. Examples of these are *T. lewisi* in rats and *T. theileri* in cattle.

3. A third mechanism evolved by the parasites to evade immune response is to change surface antigens rapidly and repeatedly. Trypanosomes have developed antigenic variation to a very high degree and are able to survive in the bloodstream in high numbers since they are transmitted by biting tsetse flies. When antigenic changes occur, the glycoproteins in the old coat are shed and replaced by antigenically different glycoproteins. The trypanosomes possess at least one thousand variant surface glycoprotein (VSG) genes. Parasite protozoa also invade immunosuppressed individuals. Organisms that are usually maintained in a quiescent state by the immune response like the *T. gondii* can change to more active form and produce severe disease in immunosuppressed animals. For this reason, acute toxoplasmosis is seen in immunosuppressed patients for transplantations and cancer therapy.

Table 16.3 Various Methods which Parasites have Evolved to Avoid Host Defense Mechanisms

Parasite	*Habitat*	*Main host effector mechanism*	*Method of avoidance*
Trypanosoma brucei	Bloodstream	Antibody + complement	Antigenic variation
Plasmodium spp.	Hepatocyte blood cell	Antibody, cytokines	Antigenic variation
Toxoplasma gondii	Macrophage	O_2 metabolites, NO, lysosomal enzymes	Failure to trigger, inhibits fusion of lysosomes
Trypanosoma cruzi	Many cells	O_2 metabolites, NO, lysosomal enzymes	Escapes into cytoplasm, so avoiding digestion
Leishmania	Macrophage	O_2 metabolites, NO, lysosomal enzymes	O_2 burst impaired and products scavenged, avoids digestion
Trichinella spiralis	Gut, blood, muscles	Myeloid cells, Antibody + complement	Encystment in muscle development
Schistosoma mansoni	Skin, blood, lungs and portal vein	Myeloid cells, Antibody + complement	Acquisition of host antigens, blockade by antibody, soluble antigens and immune complexes, antioxidants
Wuchereria bancrofti	Lymphatics	Myeloid cells, Antibody + complement	Thick extracellular cuticle, antioxidants

16.5 IMMUNITY TO HELMINTHES

Helminthes are highly prevalent in developing countries. These are large multicellular organisms that do not multiply in large numbers in humans; they are not intracellular pathogens. Since these helminthes remain inside an animal for much of its life, these parasites are able to survive immunological attacks. Both intraspecies and interspecies competition are seen to occur. In intraspecies competition (also called concomitant immunity), the presence of adult worms in the intestines delays the development of larval stages in the tissues, is seen in tapeworm, nematodes, etc. In interspecies competition, competition between worms for habitats and nutrients may control the numbers and composition of an animal's helminthes population.

Parasite worms that infect man include trematodes or flukes (e.g., schistosomes) cestodes (e.g., tapeworms) and nematodes or roundworms (e.g., hookworms, pinworms, *Ascaris* and the filarial worms). Tapeworms and hookworms live in the gut; adult schistosomes inhabit the blood vessels, while some filarial worms live in the lymphatics (Figure 16.5).

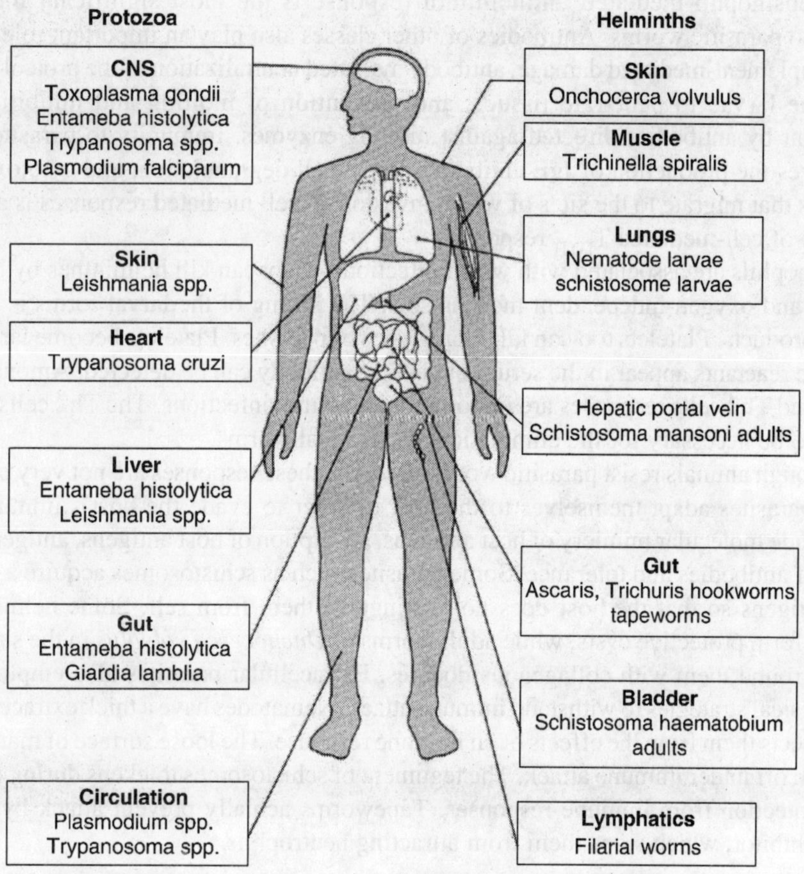

Figure 16.5 Sites of infection of medically important parasites.

The age and sex of the host influence host burden through behaviour and hormones. Children pick up infections more readily as compared to adults. In animals whose sexual cycle is seasonal, parasites tend to synchronize their reproductive cycle with that of the host. This helps the worms to spread in the new-born animals.

Worms are fully adapted obligate parasites whose survival depends on some kind of accommodation with the host. Thus, if a worm of this type causes a disease, it will be of a mild type. When parasite worms invade a host where they are not fully adapted, or are in unusually large numbers, the disease occurs in acute form.

Although helminthes are more accessible to the immune system, the immune response against them in the infected individual is very poor, may be because of their low numbers. Parasites can be present as adults in the respiratory tract and as larvae in the gastrointestinal tract. Although IgM, IgG and IgA are produced in response to worm antigens, IgE is the most significant immunoglobulin formed. Helminthes are generally tackled by antibody mediated response, which also includes ADCC.

IgE-eosinophil-mediated antihelminth response is the most significant mechanism of resistance to parasitic worms. Antibodies of other classes also play an important role towards this end by complement-mediated damage, antibody mediated neutralization of the proteolytic enzymes used by the larvae to penetrate tissues, and prevention of molting and inhibition of larval development by antibodies directed against molting enzymes. Immunity to parasite helminthes thus involves the production of IgE antibodies, mast-cell degranulation, and cytotoxic effects of eosinophils that migrate to the sites of worm invasion. T-cell mediated responses is also involved in the form of cell-mediated T_{DTH} response.

Eosinophils are associated with worm infections. They can kill helminthes by both oxygen-dependent and oxygen-independent mechanisms. The killing of the larval forms is enhanced by mast-cell products. Platelets too can kill many types of parasites. Platelets become larvicidal when acute phase reactants appear in the serum but before antibody can be detected. Among the T-cells, both Th1 and Th2 cells responses are important in helminth infections. The Th2 cells have clearly come out to be necessary for the elimination of intestinal worms.

Although animals resist parasitic worm infection, these responses are not very effective. The helminth parasites adapt themselves to the host in order to evade the host's immune response. These include molecular mimicry of host antigens, adsorption of host antigens, antigenic variation, blocking of antibodies and tolerance. Some parasites such as schistosomes acquire a surface layer of host antigens so that the host does not distinguish them from self. Some helmithes such as *T. spiralis* form protective cysts, while adult worms of *Onchocerca volvolus* in the skin induce the host to surround them with collagenous nodules. Extracellular parasites also employ some very simple physical strategies to withstand immune attack. Nematodes have a thick extracellular cuticle, which protects them from the effects of an immune response. The loose surface of many nematodes may slough off under immune attack. The tegument of schistosomes thickens during maturation to offer it protection from immune responses. Tapeworms actually prevent attack by secreting an elastase inhibitor, which stops them from attracting neutrophils.

MULTIPLE-CHOIÇE QUESTIONS

1. Antibodies can protect against viral diseases through:
   (a) Release of antibiotics from B-cells
   (b) Blocking viral adsorption to cells
   (c) T-cell derived viricidal proteins
   (d) Viricidal activity of histamine

2. Bacterial cell walls cause the activation of the:
   (a) kinin pathway
   (b) classical complement pathway
   (c) alternative complement pathway
   (d) tyrosine kinase pathway
3. Major changes in the antigenic structure of influenza viruses are called:
   (a) antigenic shift
   (b) antigenic drift
   (c) attenuation
   (d) signal transduction
4. The most important cells involved in the destruction of virus-infected cells are:
   (a) B-cells
   (b) NK cells
   (c) Macrophages
   (d) Cytotoxic T-cells
5. The most potent bactericidal factor produced by activated macrophages is:
   (a) nitric oxide
   (b) lysozyme
   (c) complement
   (d) transferrin

REVIEW QUESTIONS

16.1 What is the mode of viral infection? How does an immune response develop against a viral infection?

16.2 Discuss the strategy of viral immune evasion.

16.3 How does a body respond to bacterial invasion?

16.4 How do bacteria evade immune responses?

16.5 What are the different categories of fungal infection?

16.6 Describe the mechanisms that protozoans resort to evade the host's immune response.

16.7 Give an account of immune response against helminth parasites.

ANSWERS TO MULTIPLE-CHOICE QUESTIONS

CHAPTER 1

1. (c) 2. (d) 3. (c) 4. (d) 5. (a) 6. (b) 7. (d)

CHAPTER 2

1. (c) 2. (a) 3. (c)

CHAPTER 3

1. (d) 2. (c) 3. (a) 4. (b) 5. (b) 6. (a)

CHAPTER 4

1. (c) 2. (b) 3. (a) 4. (c) 5. (c) 6. (c) 7. (a)

CHAPTER 5

1. (c) 2. (d) 3. (a)

CHAPTER 6

1. (d) 2. (b) 3. (b) 4. (b)

CHAPTER 7

1. (b) 2. (b) 3. (d) 4. (a) 5. (c) 6. (a) 7. (b)

CHAPTER 8

1. (c) 2. (a) 3. (c) 4. (c)

CHAPTER 9

1. (c) 2. (a) 3. (c) 4. (d) 5. (b) 6. (a) 7. (d)

CHAPTER 10

1. (d) 2. (b) 3. (c) 4. (a) 5. (b) 6. (c) 7. (d)

CHAPTER 11

1. (b) 2. (d) 3. (c) 4. (a)

CHAPTER 12

1. (d) 2. (c) 3. (a) 4. (c) 5. (a) 6. (d)

CHAPTER 13

1. (b) 2. (b) 3. (a) 4. (b) 5. (c) 6. (d) 7. (c)
8. (a) 9. (c)

CHAPTER 14

1. (b) 2. (c) 3. (c) 4. (d) 5. (b) 6. (c) 7. (a)
8. (b)

CHAPTER 15

1. (c) 2. (b) 3. (b) 4. (c) 5. (b)

CHAPTER 16

1. (b) 2. (c) 3. (a) 4. (d) 5. (a)

INDEX